Arthur W. Djbdahl

OPTICS

Solutions Manual

K. D. Möller

Professor of Physics
Fairleigh Dickinson University
Teaneck, New Jersey

University Science Books
Mill Valley, California

Preface

This Solutions Manual is meant to accompany the textbook Optics and contains detailed solutions to each of the end-of-chapter problems in the text.

It is important to the comprehension of the subject matter of the text to work many problems, and the detailed solutions given here can serve as a valuable aid for that purpose if this Manual is used correctly. That is, it is not a substitute for working the problems and should not be consulted before a sincere effort has been made to work any problem.

In addition to containing the solutions to all the problems, this manual includes some derivations that are not found in the textbook.

The availability of solved problems may help applications-oriented readers find solutions to their own particular problems more quickly by analogy.

K. D. Möller

1 Geometrical Optics

1. *Prism.* Consider a prism of apex angle A and refractive index n. Calculate the dependence of the angle of deviation δ as a function of n and the angle of incidence θ_1.

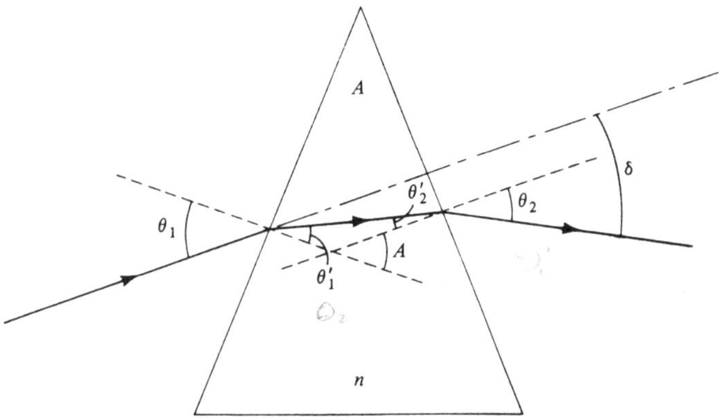

<u>Solution</u>

We have $\theta_1' + \theta_2' = A$

$$\delta = \theta_1 - \theta_1' + \theta_2 - \theta_2'$$

$$\delta = \theta_1 + \theta_2 - (\theta_1' + \theta_2') = \theta_1 + \theta_2 - A$$

We have $\sin\theta_1 = n\sin\theta_1'$

$\sin\theta_2 = n\sin\theta_2'$

$$\cos\theta_1' = \sqrt{1-\sin^2\theta_1'} = \sqrt{1-\sin^2\theta_1/n^2}$$

and $\delta = \theta_1 + \sin^{-1}[n\sin\theta_2'] - A$

$$= \theta_1 + \sin^{-1}[n\sin(A-\theta_1')] - A$$

$$= \theta_1 + \sin^{-1}[n(\sin A\cos\theta_1' - \cos A\sin\theta_1')] - A$$

$$= \theta_1 + \sin^{-1}[n(\sin A\sqrt{1-\sin^2\theta_1/n^2} - (\cos A)\frac{1}{n}\sin\theta_1)] - A$$

$$\delta = \theta_1 + \sin^{-1}[(\sin A)\sqrt{n^2-\sin^2\theta_1} - \cos A\sin\theta_1] - A$$

2. *Image formation of a real object for a single thin biconvex lens.*
   a. Fill in the following table, assuming the object is placed as indicated in column x_o.
   b. Make sketches for the five cases.

Solution

One lens system, image formation of a real object

(a)

x_o	x_i	magnification m	image real	image virtual	image enlarged	image reduced
(A) $-\infty$	f	0	✓			✓
(B) -3f	1.5f	- 1/2	✓			✓
(C) -1.5f	3f	- 2	✓		✓	
(D) -f	$+\infty$	$-\infty$		✓	✓	
(E) -0.5f	- f	2		✓	✓	

(b)

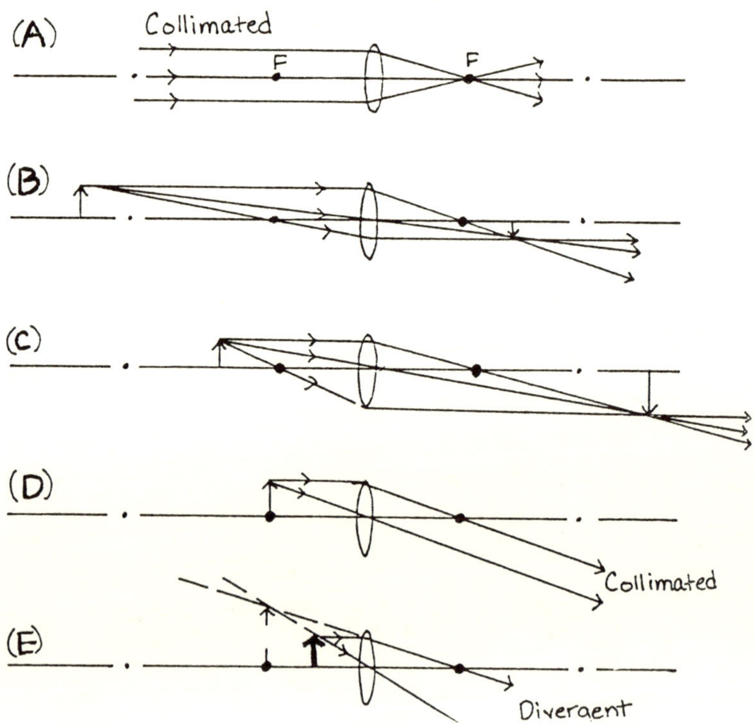

3. *Negative lens.* Consider a negative lens with $f = -5$ cm. The object is at $x_o = -10$ cm.
   a. Find x_i.
   b. Make a sketch.

Solution

(a) $\frac{1}{10} + \frac{1}{x_i} = \frac{1}{-5}$ $x_i = -3.3\text{cm}$

(b)

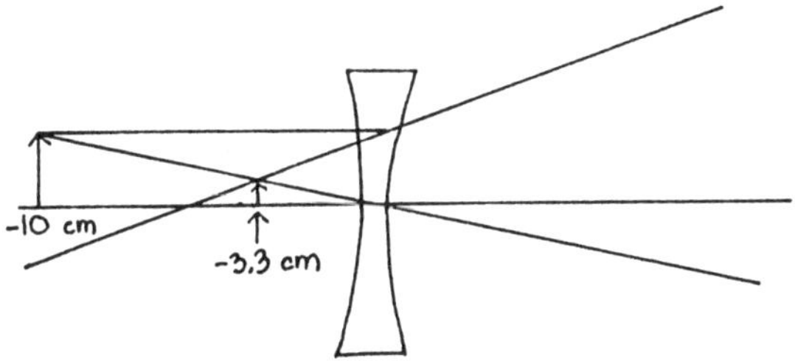

4. *System of two thin lenses.* The first lens produces a real image, and the second lens is placed so far away that the real image serves as a real object.
   a. Make a sketch.
   b. Calculate the final image position for the case where $x_{o1} = -30$ cm, $f_1 = 10$ cm, $f_2 = 15$ cm. Assume the distance between the lenses to be $d = 45$ cm.
   c. Find the magnification.

Solution

Two lenses system.

(a)

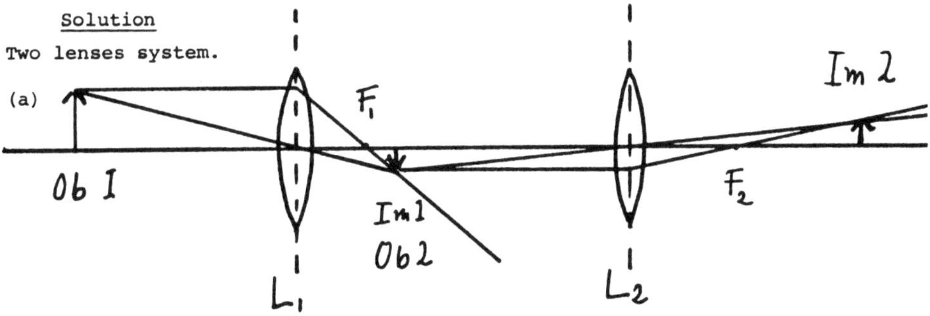

(b) $\frac{1}{-x_{o1}} + \frac{1}{x_{i1}} = \frac{1}{f_1} \Rightarrow \frac{1}{-(-30)} + \frac{1}{x_{i1}} = \frac{1}{10}$

$\Rightarrow x_{i1} = 15\text{cm}$

$\Rightarrow x_{o2} = -(45 - 15) = -30\text{cm}$

$\Rightarrow \frac{1}{x_{i2}} = \frac{1}{15} + \frac{1}{-30} \Rightarrow x_{i2} = 30\text{cm}$

(c) $m_{total} = m_1 m_2 = (\frac{x_{i1}}{x_{o1}})(\frac{x_{i2}}{x_{02}}) = (\frac{15}{-30})(\frac{30}{-30})$

$m_1 = -0.5, \quad m_2 = -1$

$m_{total} = (\frac{-1}{2})(-1) = \frac{1}{2}$

Inverted images have negative magnification.

5. *Two thin lenses in the magnifier configuration.* The first lens produces a virtual image. This image serves as the real object for the second lens (on account of its position). The focal length and distance from the object with respect to the second lens are chosen in such a way that the final image is real.

a. Make a sketch and calculate the final image location for $x_{o1} = -6$ cm, $f_1 = 12$ cm, $f_2 = 10$ cm, and a distance between the lenses of 8 cm.

b. Find the final magnification $m_{total} = m_1 \cdot m_2$.

Solution

(a)

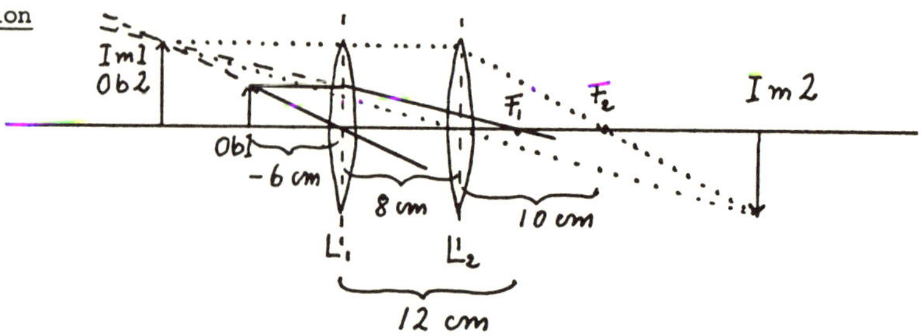

For the first lens:

$$\frac{1}{-x_{o1}} + \frac{1}{x_{i1}} = \frac{1}{f_1} \rightarrow \frac{1}{x_{i1}} = \frac{1}{12} + \frac{1}{-6} \rightarrow x_{i1} = -12\text{cm}$$

Likewise for lens L_2:

$$\frac{1}{x_{i2}} = \frac{1}{10} + \frac{1}{-(12+8)} \rightarrow x_{i2} = 20\text{cm}$$

(b)

$$m_{total} = m_1 m_2 = \left(\frac{x_{i1}}{x_{o1}}\right)\left(\frac{x_{i2}}{x_{o2}}\right) = \left(\frac{-12}{-6}\right)\left(\frac{20}{-20}\right)$$

$$= (2)\ (-1) = -2$$

6. *Two thin lenses in the "near-point configuration" of the magnifier.* If the second lens is the eye, there is a limit as to how small x_{o2} may be chosen. This distance is called the near point and is about 25 cm.

a. Make a sketch.

b. If $d = 8$ cm, $f_1 = 12$ cm, and $x_{o2} = -25$ cm, find x_{o1}.

c. If $x_{i2} = 2$ cm, what value for f_2 must we choose?

d. For this case, the angular magnification MP is given as $MP = 25/f_m + 1$. Calculate MP and give it as MPx. (See section 5C.)

Solution

(a) See Fig. 1.29 in the text for the magnifier, L_1, and eye, L_2.

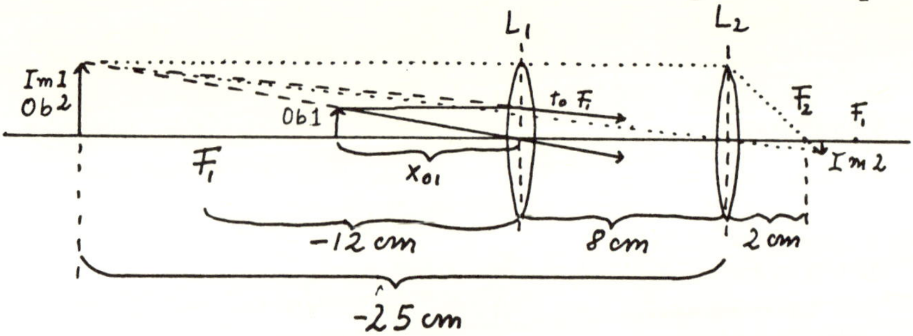

(b) $\frac{1}{-x_{01}} + \frac{1}{x_{i1}} = \frac{1}{f} \rightarrow \frac{1}{-x_{o1}} = \frac{1}{12} - \frac{1}{(8-25)} \rightarrow x_{o1} = -7.0\text{cm}$

(c) $\frac{1}{-x_{02}} + \frac{1}{x_{i2}} = \frac{1}{f_2} \rightarrow \frac{1}{f_2} = \frac{1}{25} + \frac{1}{2} \rightarrow f_2 = 1.85\text{cm}$

(d) From Eq. (1.37) one has $MP = [1 + 25/f_m]$

$MP = 1 + \frac{25}{12} = 3.08$, we have 2.08x

7. *Two thin lenses, magnifier and eye, with the first image at* ∞. Consider the eye looking at a virtual image of the first lens (magnifier) at $-\infty$. Assume $f_1 = 12$ cm, $d = 8$ cm, and $x_{i2} = 2$ cm.
   a. Where is the object to be placed; that is, find x_{o1}.
   b. Make a sketch.
   c. What is the value of f_2?
   d. Find the angular magnification MP.

Solution

Two thin lenses, magnifier and eye, with the first image at $-\infty$.

(a) $\frac{1}{-x_{c1}} + \frac{1}{x_{i1}} = \frac{1}{f_1} \Rightarrow -x_{o1} = f_1 \Rightarrow x_{o1} = -12\text{cm}$

(b)

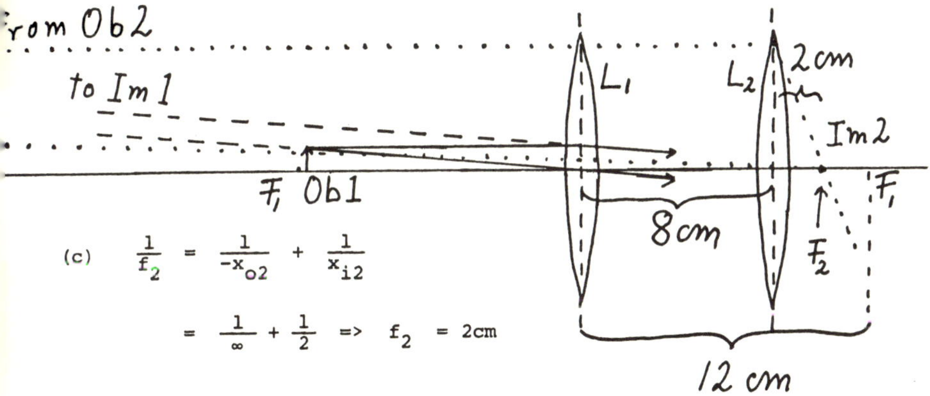

(c) $\frac{1}{f_2} = \frac{1}{-x_{o2}} + \frac{1}{x_{i2}}$

$= \frac{1}{\infty} + \frac{1}{2} \Rightarrow f_2 = 2\text{cm}$

(d) We have:

$MP \doteq \frac{25}{f_1} = \frac{25}{12} = 2.08$

8. *Two-lens system with the real image of the first lens located within the focal length of the second.* If the real image of the real object of the first lens is located within the focal length of the second lens, it produces a final virtual image.
   a. If $x_{o1} = -20$ cm, $f_1 = 10$ cm, $f_2 = 15$ cm, and $d = 30$ cm, make a sketch and calculate x_{i1} and x_{i2}.
   b. If $x_{o1} = -20$ cm, $f_1 = 10$ cm, $f_2 = 15$ cm, and $d = 10$ cm, make a sketch and find x_{i2}.

Solution

Two lens system with the real image of the first lens located within the focal length of the second.

(a)

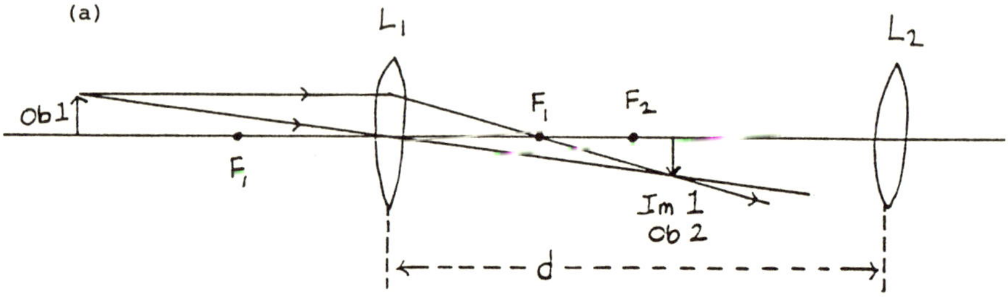

$$\frac{1}{-x_{o1}} + \frac{1}{x_{i1}} = \frac{1}{f_1} \Rightarrow \frac{1}{x_{i1}} = \frac{1}{f_1} + \frac{1}{x_{o1}} = \frac{1}{10} + \frac{1}{-20}$$

$$\Rightarrow x_{i1} = 20 \text{ cm}$$

$$\Rightarrow \frac{1}{x_{i2}} = \frac{1}{f_2} + \frac{1}{x_{o2}} = \frac{1}{15} + \frac{1}{-(30-20)} = \frac{1}{15} - \frac{1}{10}$$

$$\Rightarrow x_{i2} = -30\text{cm}$$

(b)

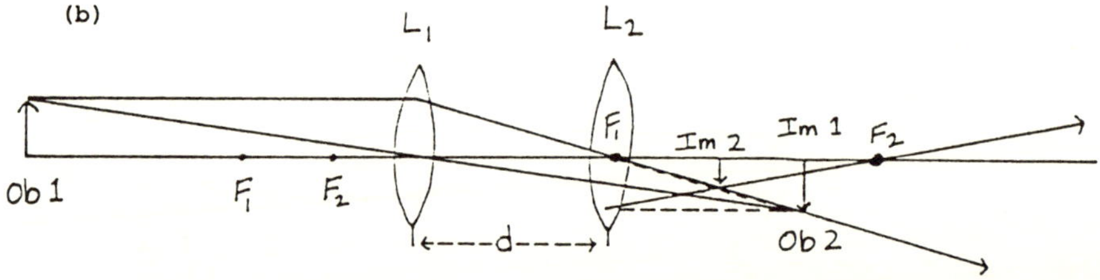

$$\frac{1}{-x_{o1}} + \frac{1}{x_{i1}} = \frac{1}{f_1} \Rightarrow x_{i1} = 20\text{cm}$$

$$\frac{1}{-x_{o2}} + \frac{1}{x_{i2}} = \frac{1}{f_2} \Rightarrow \frac{1}{-(10)} + \frac{1}{x_{i2}} = \frac{1}{15} \Rightarrow x_{i2} = 6\text{cm}$$

9. *Laser beam expander.* A laser beam of diameter 2 mm should be expanded to a beam of 20 mm.
   a. A biconvex and a biconcave lens should be used. A biconvex lens L_2 of diameter 30 mm and focal length $f_2 = 50$ mm is available. What is the focal length of the biconcave lens L_1? How far to the left of the biconvex lens is the biconcave lens to be placed?
   b. Using two biconvex lenses, one with $f_2 = 50$ mm, the other $f_1 = 5$ mm, what is the distance between the two lenses?

Solution

(a)

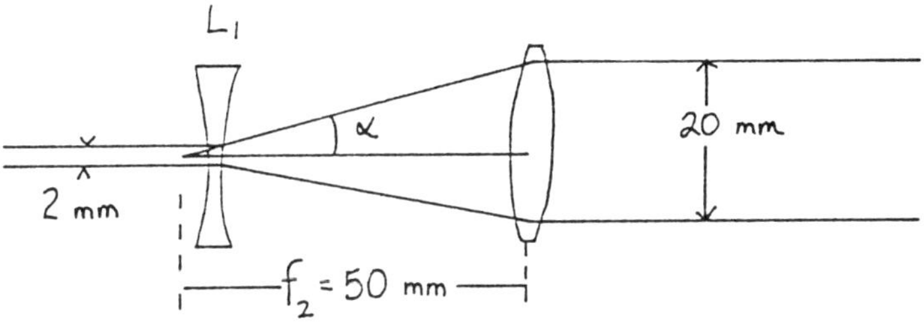

The angle α is α = arctan $\frac{10}{50}$ = 11.3°

The focal length is $f_1 = -\frac{1\text{mm}}{\tan\alpha} = -5\text{mm}$

L_1 is placed 45mm to the left of L_2.

(b)

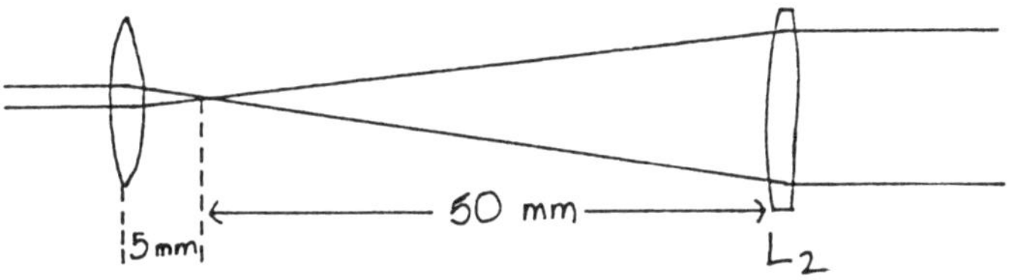

f_1 = +5mm and positioned 55mm to the left of L_2

10. *Three-lens system; microscope.* In the microscope, the objective lens forms a magnified real image. Then a magnifier configuration is applied and forms a real final image on the retina.
   a. Find the final image location x_{i3} for $x_{o1} = -2.5$ cm, $f_1 = 2$ cm, $f_m = 6$ cm, and $f_3 = 2$ cm. Assume that the distance between the first and second lens is 16 cm and that the distance between the second and third lens is 1 cm. Give x_{i1}, x_{i2}, and x_{i3}, and make a sketch (see Section 5D).
   b. The magnification M of the system is given as $M = -(16/f_1)(25/f_2)$. Calculate M.

Solution

(a)

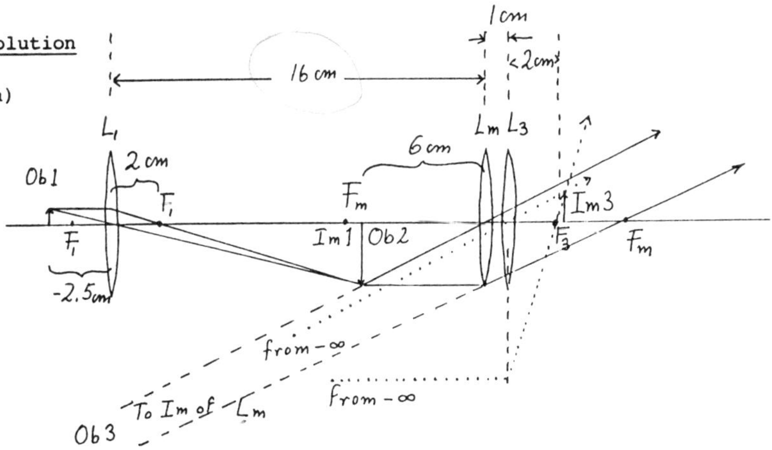

$$\frac{1}{-x_{o1}} + \frac{1}{x_{i1}} = \frac{1}{f_1} \Rightarrow \frac{1}{x_{i1}} = \frac{1}{f_1} + \frac{1}{x_{o1}} = \frac{1}{2} + \frac{1}{-2.5} \Rightarrow x_{i1} = 10\text{cm}$$

$$\frac{1}{-x_{o2}} + \frac{1}{x_{i2}} = \frac{1}{f_2} \Rightarrow \frac{1}{x_{i2}} = \frac{1}{f_2} + \frac{1}{x_{o2}} = \frac{1}{6} + \frac{1}{-6} \Rightarrow x_{i2} = -\infty$$

where the rays from lens 2 are divirging and that puts I_2 to the left of L_2 and so x_{i2} is at $-\infty$ and not $+\infty$.

$$\frac{1}{-x_{o3}} + \frac{1}{x_{i3}} = \frac{1}{f_3} \Rightarrow \frac{1}{x_{i3}} = \frac{1}{f_3} + \frac{1}{x_{o3}} = \frac{1}{2} + \frac{1}{-\infty} \Rightarrow x_{i3} = 2\text{cm}$$

(b)

$$m = -\left(\frac{16}{f_1}\right)\left(\frac{25}{f_2}\right) = -\left(\frac{16}{2}\right)\left(\frac{25}{6}\right) = -33.3$$

11. *Telescope.* The telescope is similar to the microscope. But the objective lens forms a real image from a faraway object. This object is looked at with a magnifier configuration. Assume that for the magnifier configuration, the image of this object is at $-\infty$.
   a. For $f_1 = 30$ cm, $f_2 = 6$ cm, $f_3 = 2$ cm, $d_{23} = 2$ cm, $x_{o1} = -\infty$, and $x_{i2} = -\infty$, give x_{i1}, x_{o2}, and x_{i3}, and make a sketch.
   b. What is the distance between lenses 1 and 2?
   c. The angular magnification is defined as $MP = f_1/f_2$. Give the value of MP.

Solution

Telescope

(a) For the objective lens

$$\frac{1}{-x_{o1}} + \frac{1}{x_{i1}} = \frac{1}{f_i} \Rightarrow 0 + \frac{1}{x_{i1}} = \frac{1}{30} \Rightarrow x_{i1} = 30\text{cm}$$

The drawing is not to scale:

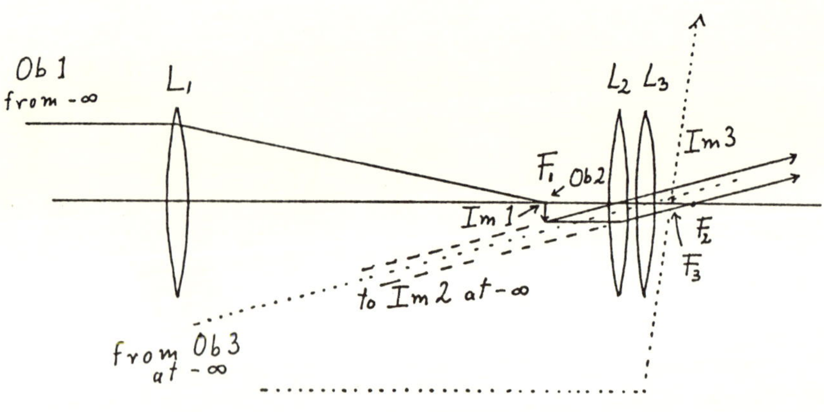

$$\frac{1}{-x_{o2}} + \frac{1}{x_{i2}} = \frac{1}{f_2} \Rightarrow \frac{1}{-x_{o2}} + 0 = \frac{1}{f_2} \Rightarrow x_{o2} = -6\text{cm}$$

$$\frac{1}{-x_{o3}} + \frac{1}{x_{i3}} = \frac{1}{f_3} \Rightarrow 0 + \frac{1}{x_{i3}} = \frac{1}{f_3} \Rightarrow x_{i3} = 2\text{cm}$$

(b) $d_{12} \equiv f_1 + f_2 = 30 + 6 = 36\text{cm}$

(c) $m = -f_1/f_2 = -30/6 = -5$

12. *Refraction and translation matrices.* We considered two glass lenses formed by interchanging the order of the refracting surfaces.

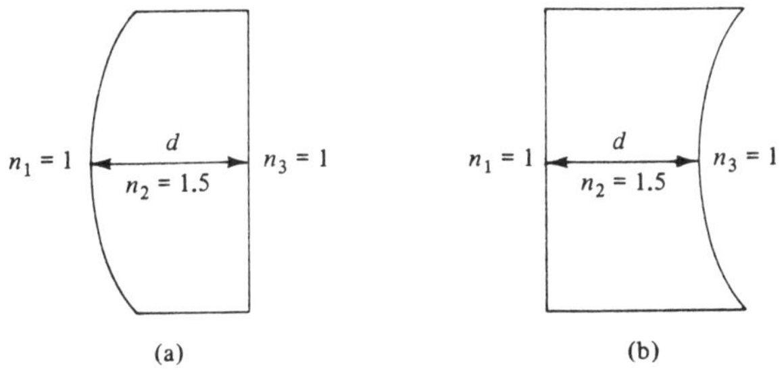

a. For the lens with the convex surface to the left, obtain the matrices describing refraction at both surfaces and translation through the lens. Take the refractive index of the glass to be 1.5. Evaluate the product of the matrices and show that the result is

$$\begin{pmatrix} 1 - \dfrac{d}{3r} & \dfrac{2}{3}d \\ -\dfrac{1}{2r} & 1 \end{pmatrix}$$

b. Repeat the calculation for the lens with the plane surface to the left. Show that the result is now

$$\begin{pmatrix} 1 & \dfrac{2}{3}d \\ \dfrac{1}{2r} & 1 + \dfrac{d}{3r} \end{pmatrix}$$

Solution

Refraction and translational matrices

(a)

We have with r > 0 for convex surfaces:

$$R_{12} = \begin{pmatrix} 1 & 0 \\ -\frac{1}{r}\left(\frac{n_2-n_1}{n_2}\right) & \frac{n_1}{n_2} \end{pmatrix} = \begin{pmatrix} 1 & 0 \\ \frac{1}{-r}\left(\frac{1.5-1.0}{1.5}\right) & \frac{1}{1.5} \end{pmatrix}$$

$$= \begin{pmatrix} 1 & 0 \\ \frac{1}{-3r} & 2/3 \end{pmatrix}$$

$$R_{23} = \begin{pmatrix} 1 & 0 \\ 0 & \frac{1.5}{1} \end{pmatrix} \qquad \text{where } r \to \infty$$

$$T = \begin{pmatrix} 1 & d \\ 0 & 1 \end{pmatrix}$$

Multiplying the matrices:

$$I_3 = \begin{pmatrix} 1 & 0 \\ 0 & 1.5 \end{pmatrix} \begin{pmatrix} 1 & d \\ 0 & 1 \end{pmatrix} \begin{pmatrix} 1 & 0 \\ -\frac{1}{3r} & 2/3 \end{pmatrix} I_1$$

$$= \begin{pmatrix} 1 & 0 \\ 0 & 1.5 \end{pmatrix} \begin{pmatrix} \left(1-\frac{d}{3r}\right) & (2/3)d \\ -\frac{1}{3r} & \frac{2}{3} \end{pmatrix} I_1 = \begin{pmatrix} 1-\frac{d}{3r} & \frac{2}{3}d \\ -\frac{1}{2r} & 1 \end{pmatrix} I_1$$

(b)

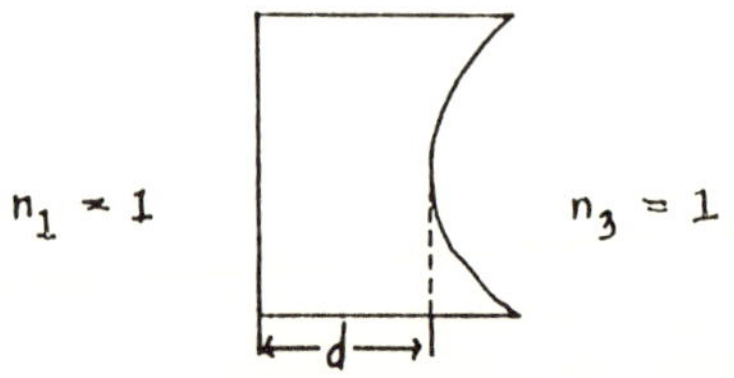

$$I_3 = R_{23}\, T\, R_{12}\, I_1$$

r > 0 for convex surfaces (viewed from the left)

$$R_{12} = \begin{pmatrix} 1 & 0 \\ 0 & 2/3 \end{pmatrix}$$

$$T = \begin{pmatrix} 1 & d \\ 0 & 1 \end{pmatrix}$$

$$R_{23} = \begin{pmatrix} 1 & 0 \\ \frac{1}{2r} & 3/2 \end{pmatrix}$$

$$I_3 = \begin{pmatrix} 1 & 0 \\ \frac{1}{2r} & 3/2 \end{pmatrix} \begin{pmatrix} 1 & d \\ 0 & 1 \end{pmatrix} \begin{pmatrix} 1 & 0 \\ 0 & 2/3 \end{pmatrix} I_1 = \begin{pmatrix} 1 & 0 \\ \frac{1}{2r} & \frac{3}{2} \end{pmatrix} \begin{pmatrix} 1 & \frac{2}{3}d \\ 0 & \frac{2}{3} \end{pmatrix} I_1$$

$$= \begin{pmatrix} 1 & \frac{2}{3}d \\ \frac{1}{2r} & \frac{d}{3r} + 1 \end{pmatrix} I_1$$

13. *Comparison of a thick and thin lens.* Consider a thick, biconvex lens in air with radii of curvature $r_1 = 20$ cm and $r_2 = -10$ cm, index of refraction 1.5, and thickness 5 cm. An object is located 8 cm to the left of the first surface.

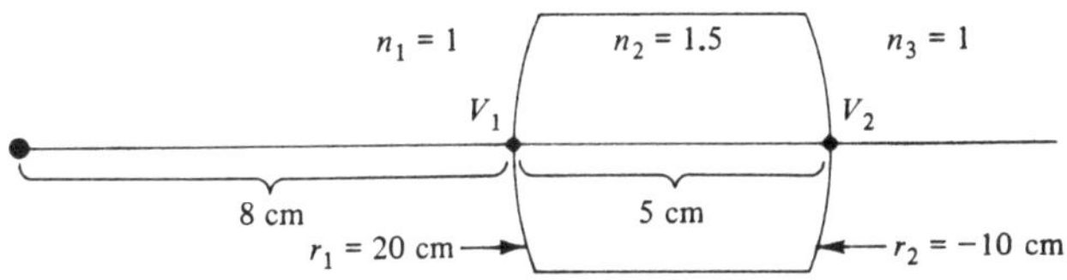

a. Calculate the image distance from the second surface by successively using the single-surface refraction equation for each surface.
b. Apply the thin-lens equation to this lens. First calculate the focal length f, using

$$\frac{1}{f} = (n_2 - 1)\left(\frac{1}{r_1} - \frac{1}{r_2}\right)$$

and then determine the image distance from the second surface. (Now all distances should be measured from the center of the lens.)
c. Treat the original thick-lens using the matrix method.

Solution

(a) First surface, convex

$$\frac{n_1}{-x_{o1}} + \frac{n_2}{x_{i1}} = \frac{n_2 - n_1}{r_1}$$

$$\frac{1}{-(-8)} + \frac{1.5}{x_i} = \frac{1.5 - 1.0}{20} \Rightarrow x_{i1} = -15\text{cm}$$

=> 1st image is 20cm to the left of the 2nd surface

For the 2nd surface, concave

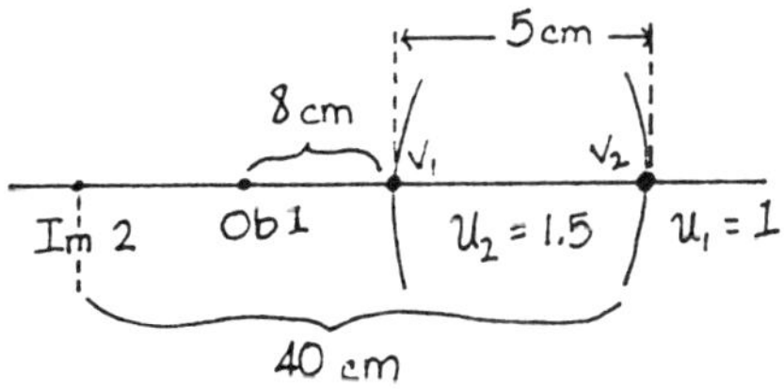

$$\frac{n_2}{-x_{o2}} + \frac{n_3}{x_{i2}} = \frac{n_3 - n_2}{r_2}$$

$$\frac{1.5}{-(-20)} + \frac{1}{x_{i2}} = \frac{1 - 1.5}{-10} \Rightarrow$$

$x_{i2} = -40$cm
measured from V_2
to the left

(b) For a thin lens:

$$\frac{1}{-x_o} + \frac{1}{x_i} = \frac{1}{f} = \frac{n_2 - 1}{r_1} + \frac{1 - n_2}{r_2} = (n_2 - 1)\ (\frac{r_2 - r_1}{r_1 r_2})$$

where for a thin lens ($d \to 0$) $x_i' - x_i$ and $r \to r_1$ with $r' \to r_2$. The 1st surface is convex => $r_1 > 0$; the 2nd surface is concave => $r_2 < 0$. See text just before Eq. (1.11).

$$f = \frac{r_1 r_2}{(n_2 - 1)(r_2 - r_1)} = \frac{20(-10)}{(1.5 - 1.0)[-10 - 20]} = 13.33\text{cm}$$

$$\Rightarrow \frac{1}{-(-8 - 2.5)} + \frac{1}{x_i} = \frac{1}{13.3} \Rightarrow \{x_i = -49.41$$ measured from the center of the lens}

(or $x_i = -50 - 2.5 = -52.5$ cm measured from V_2 to the left.)

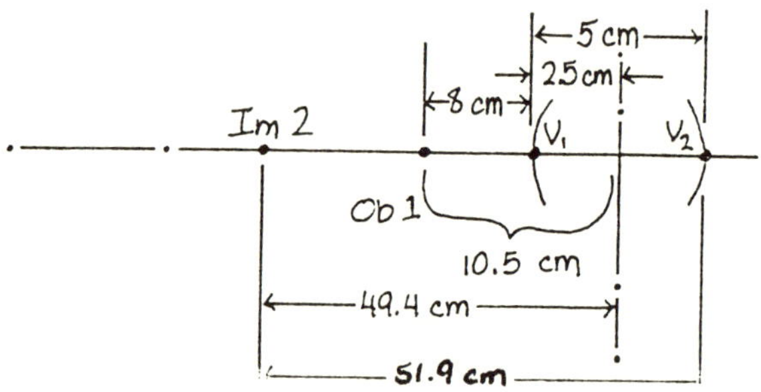

(c) For a thick lens we first locate h' and h.

$$P_{12} = -\frac{1}{r_1}(\frac{n_2 - n_1}{n_2}) = \frac{1}{-20}(\frac{1.5 - 1}{1.5}) = -\frac{1}{60} = -0.0167$$

$$P_{23} = -\frac{1}{r_2}(\frac{n_3 - n_2}{n_3}) = \frac{-1}{-10}(\frac{1 - 1.5}{1}) = -\frac{1}{20} = -0.0500$$

$$P = -\frac{1}{f} = P_{23} + dP_{12}P_{23} + \frac{n_2}{n_3} P_{12}$$

$$= [-0.0500 + 5(-0.0167)(-0.0500) + \frac{1.5}{1}(-0.0167)]$$

$$= -0.0708 \Rightarrow f = 14.1\text{cm}$$

$$h = -\frac{d}{P} P_{23} \frac{n_1}{n_2} = -\frac{5}{-0.0708}(-0.0500)\frac{1}{1.5} = -2.35\text{cm}$$

$$h' = -\frac{d}{P} P_{12} = -\frac{5(-0.0167)}{-0.0708} = -1.18\text{cm}$$

(Note the negatives)

Now use $\frac{1}{-x_o} + \frac{1}{x_i} = \frac{1}{f}$

Measured from H $\frac{1}{-(-8-2.35)} + \frac{1}{x_i} = \frac{1}{14.1} \Rightarrow x_i = -38.8\text{cm}$ Measured from H'

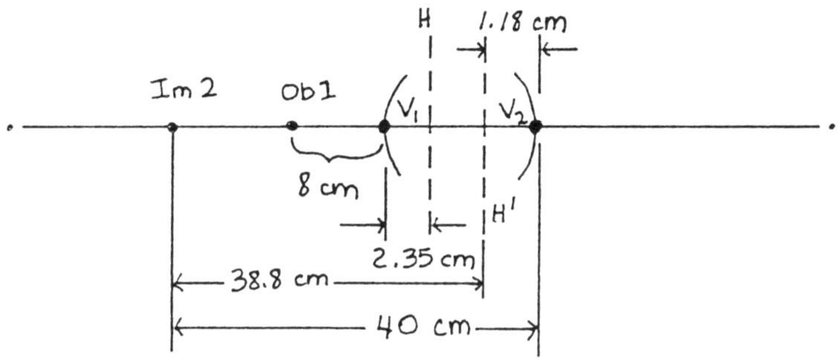

This thick lens approach (c) agrees with the calculations for one surface at a time (a) and differs markedly from the thin lens approximation (b).

14. *Biconcave lens.* Consider a biconcave lens in air with radii $r_1 = -10$ cm and $r_2 = 5$ cm and a thickness at the vertex of 3 cm. An object is 5 cm to the left of the first vertex.

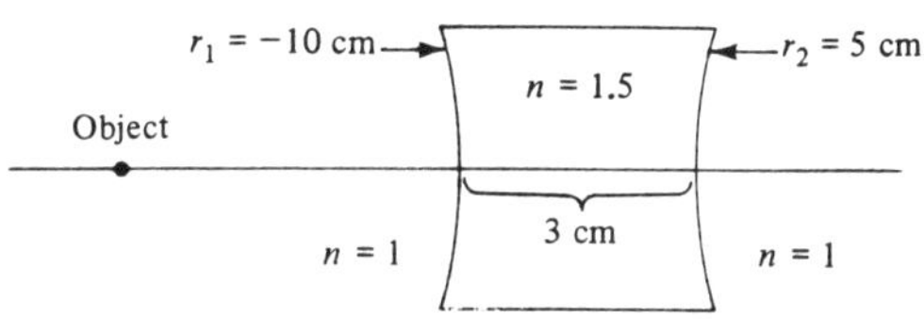

a. Using the equation for image formation

$$-\frac{n_1}{x_o} + \frac{n_2}{x_i} = \frac{n_2 - n_1}{r}$$

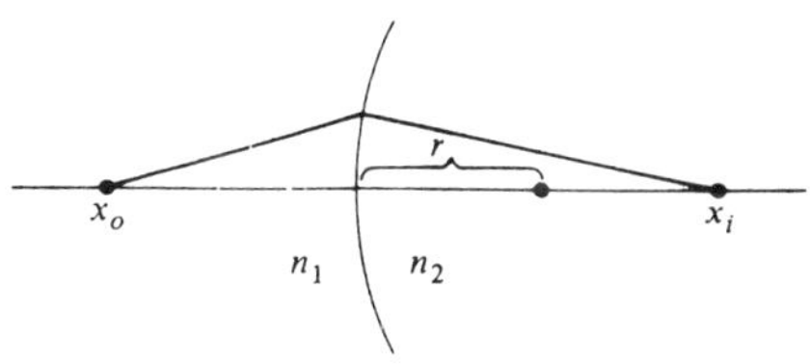

at each surface of the lens, find the final image.

b. Use the thin-lens equation to solve this problem, taking the center of the lens as a reference point.

c. Treat the thick, biconcave lens using the matrix method.

Solution

Biconcave Lens

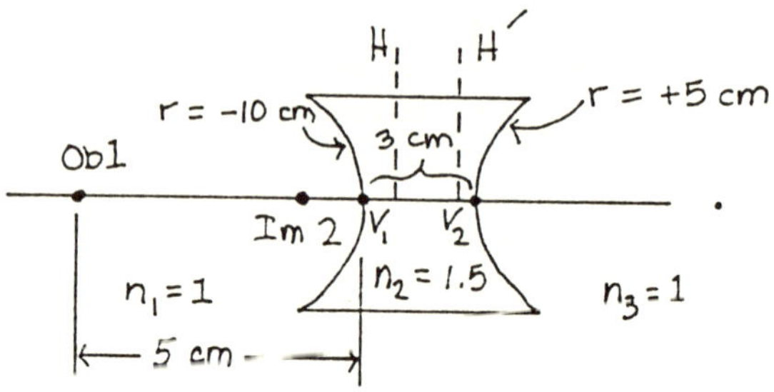

(a) Two surfaces, one at a time

$$\text{Eq. (1.11)} \Rightarrow \frac{n_1}{-x_{o1}} + \frac{n_2}{x_{i_1}} = \frac{n_2-n_1}{r_1} \Rightarrow \frac{1}{-(-5)} + \frac{1.5}{x_{i1}} = \frac{1.5-1.0}{-10}$$

$\Rightarrow x_{i1} = -6.0\text{cm}$ to the left of the 1st surface

at the 2nd surface

$$\frac{n_2}{-x_{o2}} + \frac{n_3}{x_{i2}} = \frac{n_3-n_2}{r_2}$$

$$\frac{1.5}{-(-6-3)} + \frac{1}{x_{i2}} = \frac{1-1.5}{5} \Rightarrow \{x_{i2} = -3.75\text{cm to the left of } V_2\}$$

(b)

$$\frac{1}{f} = (n_2-1)\left(\frac{1}{r_1} - \frac{1}{r_2}\right) = (1.5-1)\left(\frac{1}{-10} - \frac{1}{5}\right)$$

$\Rightarrow f = -6.67\text{cm}$

$$\frac{1}{-x_o} + \frac{1}{x_i} = \frac{1}{f} \Rightarrow \frac{1}{-(-5-3/2)} + \frac{1}{x_i} = \frac{1}{-6.67}$$

$\Rightarrow x_i = -3.29\text{cm}$ {measured to left from the center of the lens or

$-3.29-1.5 = -4.79\text{cm}$ to the left of V_2}

(c) Thick lens approach

$$P_{12} = -\frac{1}{r_1}\left(\frac{n_2-n_1}{n_2}\right) = \frac{1}{-10}\,\frac{(1.5-1.0)}{1.5} = \frac{1}{30} = 0.0333$$

$$P_{23} = -\frac{1}{r_2}\left(\frac{n_3-n_2}{n_3}\right) = \frac{-1}{5}\left(\frac{1.0-1.5}{1.0}\right) = \frac{1}{10} = 0.100$$

$$P = -\frac{1}{f} = P_{23} + d\,P_{12}P_{23} + \frac{n_2}{n_3}P_{12}$$

$$= 0.100 + 3(0.0333)(0.100) + \frac{1.5}{1.0}(0.0333)$$

$$= 0.160 \Rightarrow f = -6.25\text{cm}$$

$$\left\{\begin{aligned} h &= -d\,\frac{P_{23}}{P}\,\frac{n_1}{n_2} = \frac{-(3)(0.100)}{0.160}\left(\frac{1}{1.5}\right) = -\ 1.25\text{cm} \\ h' &= -d\,\frac{P_{12}}{P} = \frac{-3(0.0333)}{0.16} = -0.625\text{cm} \end{aligned}\right.$$

Note the negative signs!

$$\frac{1}{-x_o} + \frac{1}{x_i} = \frac{1}{f}$$

$$\frac{1}{-(-1.25-5)} + \frac{1}{x_i} = \frac{1}{-6.25} \Rightarrow \{x_i = -3.125\text{cm measured from H' or}$$

$(-3.125-0.625) = -3.75$cm measured to the left of V_2 in agreement with part (A).}

15. *Short focal length biconvex lens.* A quartz lens should be used for concentrating parallel light on a detector. In the wavelength region considered, quartz has $n = 2$. A biconvex lens of thickness 4 mm should be used, and the detector should be placed 2 mm to the right of the second surface. We want to find the radii of curvature.
   a. Express x_{i1} by the radius r_1 from the first surface.
   b. Express x_{o2} with x_{i1} and d, and write the imaging equation for the second surface.
   c. Assume $r_1 = -r_2$ and solve the quadratic equation for r_1.
   Make the following checks:
   d. Calculate x_{i1} from (a), using the value of r_1, and convert to x_{o2}.
   e. Calculate x_{i2} from (b) (should be 2 mm).

<u>Solution</u>

(a)

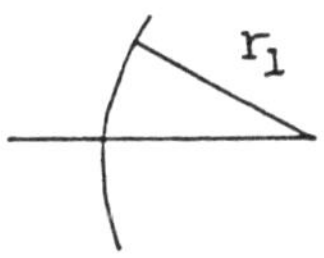

$$\frac{n_1}{-x_{o1}} + \frac{n_2}{x_{i1}} = \frac{n_2 - n_1}{r_1}$$

$x_o = -\infty$ we have $\frac{2}{x_{i1}} = \frac{2-1}{r_1} = \frac{1}{r_1}$ $\quad x_{i1} = 2\ r_1$

(b)

$-x_{o2} = 4 - x_{i1} = 4 - 2\ r_1\ ,\ \frac{2}{4-2r_1} + \frac{1}{x_{i2}} = \frac{1-2}{r_2}$

$x_{i2} = 2\text{mm}$

$$\frac{2}{4-2r_1} + \frac{1}{2} = \frac{1-2}{r_2}$$

(c)

$r_1^{\,2} - 6\ r_1 = -4$

$r_1 = 3 \pm \sqrt{9-4}$ $\quad r_1 = 5.23$ mm (for + solution)

(d)

$$x_{i1} = 2\,r_1 = 10.46$$

$$-x_{02} = 4 - x_{i1} = 4 - 2\,r_1 \qquad x_{o2} = 6.46 \text{ mm}$$

(e)

$$\frac{2}{4 - 2r_1} + \frac{1}{x_{i2}} = \frac{1-2}{r_2} \rightarrow \frac{+2}{-6.46} + \frac{1}{x_{i2}} = \frac{-1}{-5.23}$$

$$\frac{1}{x_{i\,2}} = \frac{1}{5.23} + \frac{2}{6.46} \rightarrow x_{i2} = 1.996 \text{ mm}$$

16. *Short-focal-length biconvex lens as thick lens.* In problem 15, we saw that a quartz lens of index of refraction $n = 2$ and radii of curvature $-r_1 = +r_2 = -5.23$ mm concentrates parallel light from the left onto a detector at a distance 2 mm from the second surface to the right. Check the validity of all numbers by application of the thick-lens formulas.

Solution

$$P_{12} = -\frac{1}{r_1}\,\frac{n_2 - n_1}{n_2} = -\frac{1}{r_1}\,\frac{1}{2} = -\ .0955$$

$$P_{23} = -\frac{1}{r_2}\,\frac{n_3 - n_2}{n_3} = \frac{-1}{r_1} = -\ .1912$$

$$P = P_{23} + d\cdot P_{12}\,P_{23} + \frac{n_2}{n_3}\,P_{12}$$

$$P = +\frac{2}{r_1} + 2\left(\frac{1}{r_1}\right)^2 = +\ .3824 + .0731 = -\ .3093$$

$$h' = -\,d\,\frac{P_{12}}{P} = -4\,\frac{-.0955}{-.3093} = -1.235$$

$$\frac{1}{-x_o} + \frac{1}{x_i} = \frac{1}{f}$$

$x_o = -\infty \rightarrow x_i = f = -\frac{1}{P} = 3.233$ that is x_i from H'

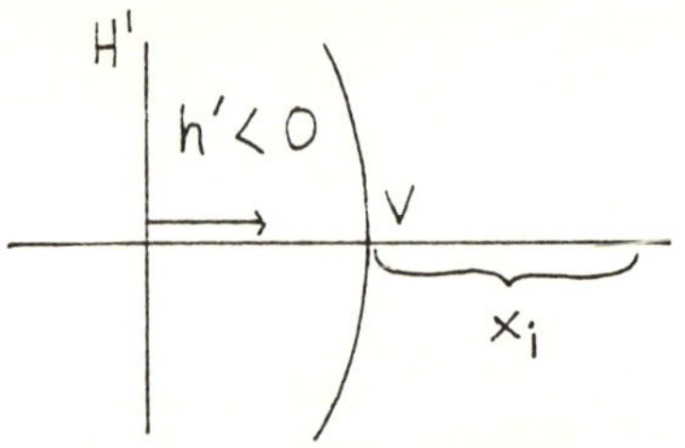

x_i from V_2 to the right is

$3.233 - 1.235 = 1.998$ mm

17. *Short-focal length planoconvex lens.* In problems 15 and 16, we considered a biconvex lens of thickness 4 mm and radii of curvature $r_1 = -r_2 = -5.23$ mm. The light was incident parallel from the left and the image position was sought for a lens with $n = 2$. Now we want to find the image of a planoconvex lens of thickness $d = 4$ mm.

a. Calculate the image position from the second plane surface.
b. Make the lens so thick that the image is on the second surface. What is the thickness d' of the lens?
c. Check the calculation of (b) with the matrix formulation of the thick lens.

Solution

(a)

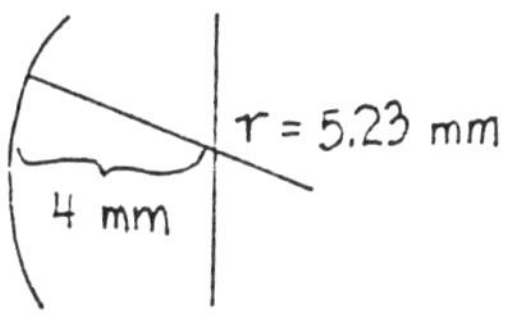

$$\frac{n_1}{-x_{o1}} + \frac{n_2}{x_{i1}} = \frac{n_2 - n_1}{r_1}$$

$$x_{o1} = -\infty$$

$$\frac{n_2}{x_{i1}} = \frac{n_2 - n_1}{r_1} \rightarrow \frac{2}{x_{i1}} = \frac{1}{r_1}$$

$$x_{i1} = 2\, r_1 \rightarrow x_{i1} = 10.46\text{mm}$$

$$-x_{o2} = 4 - x_{i1} = 4 - 10.46 = -6.46 \text{ mm}$$

$$r_2 = \infty \qquad -\frac{2}{6.46} + \frac{1}{x_{i2}} = 0 \qquad x_{i2} = \frac{6.46}{2} = 3.23 \text{ mm}$$

(b)

$$d' = x_{i1} = 10.46\text{mm}$$

(c)

$$P_{12} = -\frac{1}{r_1}\,\frac{n_2 - n_1}{n_2} = -.0955 = -\frac{1}{2r_1}$$

$$P_{23} = 0 \qquad P = 2\,P_{12} = -\frac{1}{r_1} = -.1912$$

$$h' = (-d')\left(-\frac{1}{2r_1}\right)\frac{1}{-\frac{1}{r_1}} = \frac{-d'}{2}$$

$$\frac{1}{-x_o} + \frac{1}{x_i} = \frac{1}{f} \rightarrow \frac{1}{x_i} = \frac{1}{f} \qquad x_i = f = -\frac{1}{P} = 5.23 \text{ mm}$$

This agrees with b if $d' = 10.46$mm

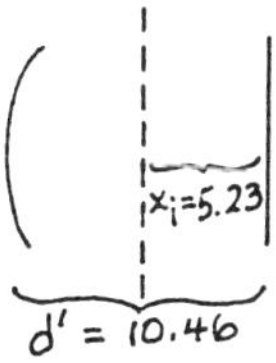

✓ 18. *Glass sphere.* Consider a glass sphere ($n = 1$) of radius 10 cm. An object is located 120 cm to the left of the sphere (measured from the first refracting surface).

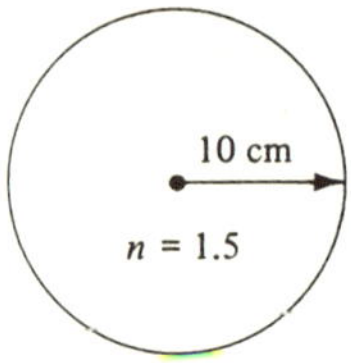

a. Determine the image point of the object, using the single-surface refraction equation at each surface.
b. Assume that the thin-lens equation can be applied. Solve for the final image point position, using the center of the sphere as a reference.
c. Use the matrix method to solve this problem and apply the formulas derived in Section 7C.
d. Determine the parameters for equal object and image distances for the sphere, using the center as a reference point and $f = nr/2(n-1)$. Make a sketch and express all distances in terms of the diameter of the sphere.

Solution

Glass Sphere

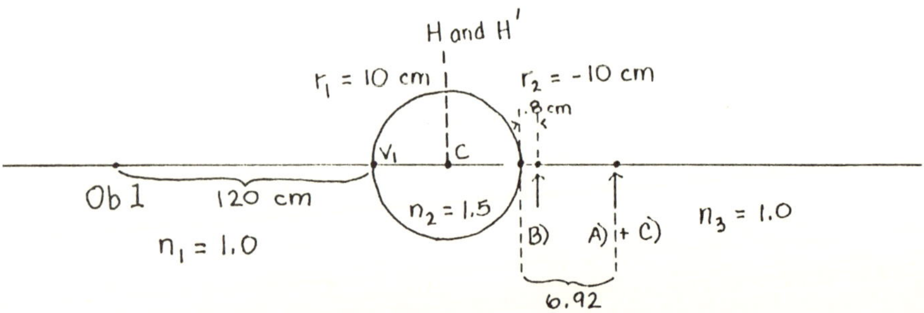

(a) one surface at a time

$$\frac{n_1}{-x_{o1}} + \frac{n_2}{x_{i1}} = \frac{1}{f_1} = \frac{n_2-n_1}{r_1} = \frac{1.5-1.0}{10} = \frac{1}{20}$$

$$\Rightarrow \frac{1}{-(-120)} + \frac{1.5}{x_{i1}} = \frac{1}{20} \Rightarrow x_{i1} = 36\text{cm}$$

2nd surface: $\frac{1}{f} = \frac{n_3-n_2}{r_2} = \frac{1.0-1.5}{-10} = \frac{1}{20}$

$\Rightarrow \quad f_2 = 20\text{cm}$

$$\frac{n_2}{-x_{o2}} + \frac{n_3}{x_{i2}} = \frac{1}{f_2} \Rightarrow \frac{1.5}{-(36-20)} + \frac{1}{x_{i2}} = \frac{1}{20}$$

$\Rightarrow \quad x_{i2} = 6.96\text{cm}$ to the right of V_2

(b) thin lens approach

$$\frac{1}{f} = (n_2-1)\left(\frac{1}{r_1} - \frac{1}{r_2}\right) = (1.5-1.0)\left(\frac{1}{10} - \frac{1}{-10}\right)$$

$\Rightarrow f = 10\text{cm}$

$$\frac{1}{-x_o} + \frac{1}{x_i} = \frac{1}{f} \Rightarrow \frac{1}{-(-120-10)} + \frac{1}{x_i} = \frac{1}{10}$$

$\Rightarrow$ x_i = 10.8cm to the right of the center of the lens or 0.8cm to the right of V_2

(c) thick lens approach

$$P_{12} = -\frac{1}{r_1}\left(\frac{n_2-n_1}{n_2}\right) = \frac{-1}{10}\left(\frac{1.5-1.0}{1.5}\right) = -0.0333 \left(= -\frac{1}{3r}\right)$$

$$P_{23} = -\frac{1}{r_2}\left(\frac{n_3-n_2}{n_3}\right) = \frac{-1}{-10}\left(\frac{1-1.5}{1}\right) = -0.0500 \left(= -\frac{1}{2r}\right)$$

$$P = -\frac{1}{f} = P_{23} + d\,P_{12}\,P_{23} + \frac{n_2}{n_3}\,P_{12}$$

$$= -0.0500 + 20\,(-0.0333)(-0.0500) + \frac{1.5}{1}\,(-0.0333)$$

$$= -0.0667 \Rightarrow f = 15\text{cm}$$

$$h = -d\,\frac{P_{23}}{P}\,\frac{n_1}{n_2} = -20\,\frac{(-0.0500)}{-0.0667}\,\frac{1}{1.5}$$

$$= -10\text{cm}$$

$$h' = -d\,\frac{P_{12}}{P} = \frac{-20\,(-0.0333)}{-0.0667}$$

$$= -10\text{cm}$$

$\Rightarrow$ H and H' are at the center of the sphere

$$\frac{1}{-x_o} + \frac{1}{x_i} = \frac{1}{f} \Rightarrow \frac{1}{-(-120-10)} + \frac{1}{x_i} = \frac{1}{15}$$

$\Rightarrow$ x_i = 17cm to the right of H' or 7cm to the right of V_2 which is the same answer for part (a).

(d) equal object and image distances

$x_o = -x$ and $x_i = +x$

$$\frac{1}{-x_o} + \frac{1}{x_i} = \frac{1}{f} \Rightarrow \frac{1}{-(-x)} + \frac{1}{x} = \frac{1}{f}$$

$\Rightarrow f = \frac{x}{2}$

Now to find f we have (Note: $d = 2r$ for the sphere)

$$P_{12} \equiv -\frac{1}{r_1}\left(\frac{n_2-1}{n_2}\right) = -\frac{1}{3r}$$

$$P_{23} = -\frac{1}{r_2}\left(\frac{1-n_2}{1}\right) = -\frac{1}{2r}$$

$$P = -\frac{1}{f} = P_{23} + d\,P_{12}\,P_{23} + \frac{n_2}{n_1}\,P_{12}$$

$$= \left\{ \frac{1}{r}(1-n_2) + 2r\left[-\frac{(n_2-1)}{rn_2}\right]\left[\frac{1}{r}(1-n_2)\right] + n_2\left[-\frac{(n_2-1)}{rn_2}\right] \right\}$$

$$= -\frac{(n_2-1)}{r} + \frac{2(n_2-1)^2}{rn_2} - \frac{(n_2-1)}{r} = \frac{-2(n_2-1)}{r} + \frac{2(n_2-1)^2}{rn_2}$$

$$= \frac{-2(n_2-1)}{r}\left[1 - \frac{n_2-1}{n_2}\right] = \frac{-2(n_2-1)}{r}\left[\frac{n_2-n_2+1}{n_2}\right]$$

$$\Rightarrow\ -\frac{1}{f} = \frac{-2(n_2-1)}{n_2 r} \Rightarrow f = \frac{n_2 r}{2(n_2-1)}$$ for the <u>sphere</u> in air

for $n_2 = 1.5$ and $r = 10\text{cm}$ $f = \dfrac{1.5(10)}{2(1.5-1.0)} = 15\text{cm}$

$x = 2f = 2(15) = 30\text{cm}$ (or: $x = \dfrac{n_2 r}{(n_2-1)} = 3r$)

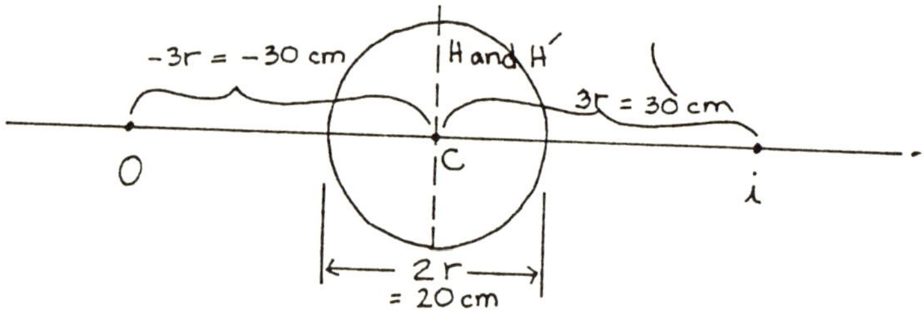

19. *Corner mirror.* Show that a corner mirror (90°) will reflect incident light through an angle of 180° regardless of the angle α between incident ray and mirror. ($\mathbf{n}_1$ is a normal.)

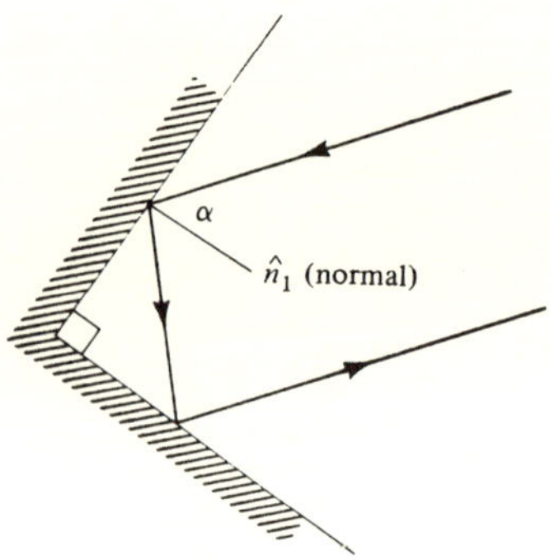

<u>Solution</u>

Reflecting Surfaces: Corner Mirror

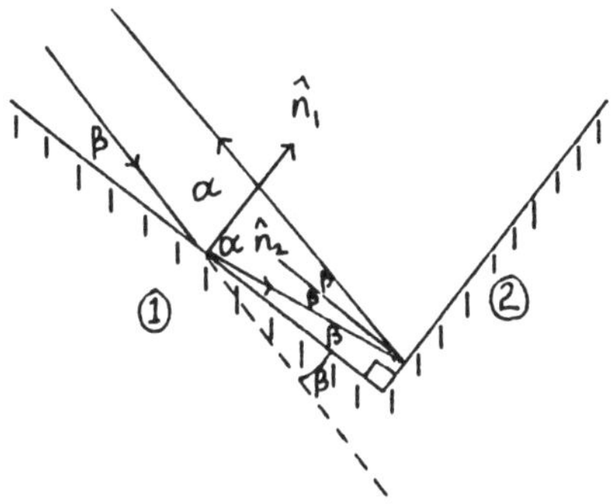

$\hat{n}_1$ and $\hat{n}_2$ are normals to two normal surfaces.

(parallel input and output rays) => [(from surface 1: angle of incidence plus angle of reflection) + (from surface 2: angle of incidence plus angle of reflection)]

$$\Rightarrow \quad \pi = [(2\alpha) + (2\beta)]$$

This is the condition for retroreflection. Noting the straight line formed by the incoming ray and the dashed line, we can write:

$$\pi = \alpha + \alpha + \beta + \beta = 2\alpha + 2\beta.$$

This expression from geometry is exactly the same for retroreflection. Since α and β are any set of angles, one has retroreflection for the corner cube.

20. *Concave mirror.* Consider a concave mirror with $r = -12$ cm. The object is at $x_o = -20$ cm.
   a. Find x_i.
   b. Make a sketch.

Solution

Consider a concave mirror with r = -12 cm

a) The object is at $x_o = -20$. Give x_i

$$\frac{1}{-20} + \frac{1}{x_i} = \frac{1}{-6} \qquad \frac{1}{x_i} = -.166 + .05$$

$$= -.116$$

$$x_i = -8.6 \text{ cm}$$

b) Make a sketch.

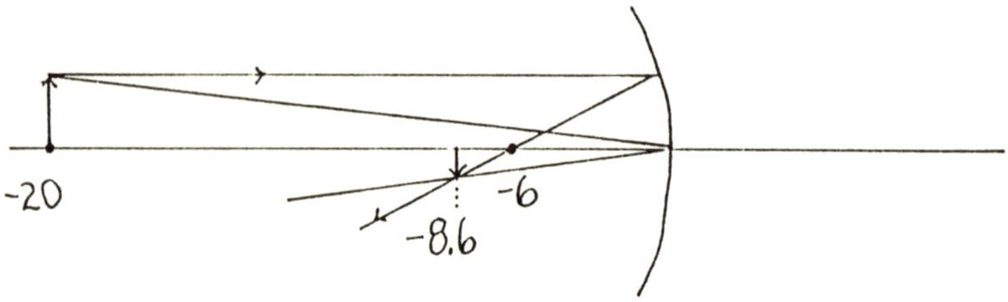

21. *Convex mirror.* Consider a convex mirror with $r = 12$ cm. The object is at $x_o = -20$ cm.
   a. Find x_i.
   b. Make a sketch.

Solution

Consider a convex mirror of r = 12 cm

a) object is at $x_o = -20$.

$$\frac{1}{-20} + \frac{1}{x_i} = \frac{1}{6} \qquad \frac{1}{x_i} = .166 + .05$$

$$x_i = + 4.6 \text{ cm}$$

b) Make a sketch.

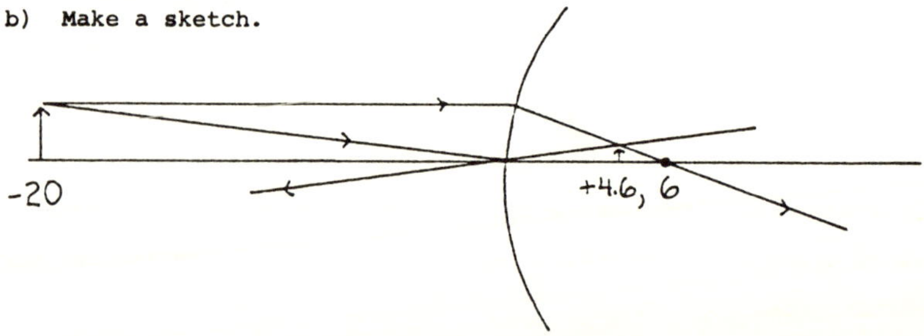

22. *Concave spherical mirror, analytical solution.* Use $r = -10$ cm and find x_i for the following:
   a. $x_o = -\infty$
   b. $|x_o| > |r|, x_o = -15$ cm
   c. $|x_o| = |r|, x_o = -10$ cm
   d. $|r| \geqq |x_o| > |f|, x_o = -8$ cm
   e. $|x_o| < |f|, x_o = -3$ cm

Solutions

(a) $x_o = -\infty \quad \frac{1}{x_o} + \frac{1}{x_i} = \frac{1}{f} = \frac{2}{r} \rightarrow \frac{1}{x_i} = \frac{1}{f}, \quad x_i = f$

$$r = -10\text{cm} \rightarrow \underline{f = -5\text{cm} = x_i}$$

(b) $|x_o|>|r|$, $x_o = -15cm$

$$\frac{1}{x_i} = \frac{1}{f} - \frac{1}{x_o} = \frac{1}{-5} - \frac{1}{-15} = \frac{1}{-7.5} \rightarrow \underline{x_i = -7.5cm}$$

(c) $|x_o|=|r|$, $x_o = -10cm$

$$\frac{1}{x_i} = \frac{1}{-5} - \frac{1}{-10} = \frac{1}{-10} \rightarrow \underline{x_i = -10cm}$$

(d) $|r|\geq|x_o|>|f|$, $x_o = -8cm$

$$\frac{1}{x_i} = \frac{1}{-5} - \frac{1}{-8} = \frac{1}{-13.3} \rightarrow \underline{x_i = -13.3cm}$$

(e) $|x_o|<|f|$, $x_o = -3cm$

$$\frac{1}{x_i} = \frac{1}{-5} - \frac{1}{-3} = \frac{1}{7.5} \rightarrow \underline{x_i = 7.5cm}$$

23. *Concave spherical mirror, graphical solution.* Using the graphical method, find the image produced by a concave spherical mirror (r negative) for the following values of object distances:
   a. $x_o = -\infty$
   b. $|x_o| > |r|$
   c. $|x_o| = |r|$
   d. $|r| \geqq |x_o| > |f|$
   e. $|x_o| < |f|$

<u>Solution</u>

Concave spherical mirror - graphical solution.

(a) $x_o = -\infty$

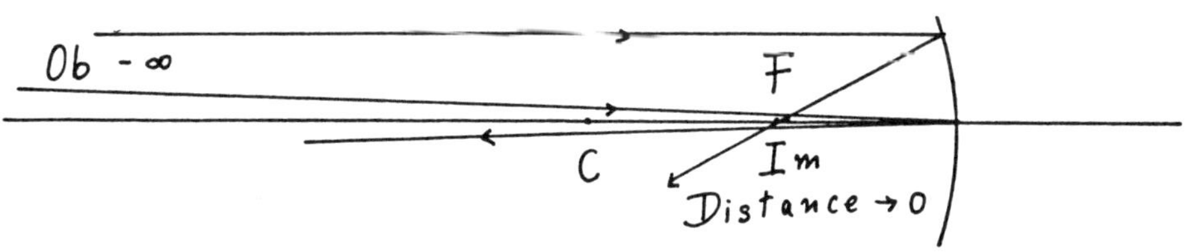

(b) $|x_o| > |r|$

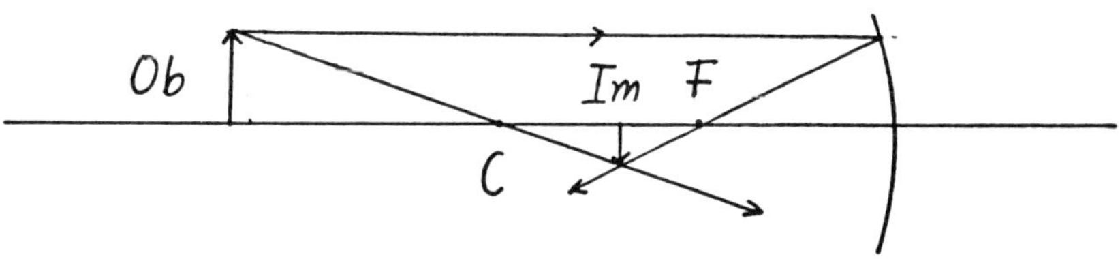

(c) $|x_o| = |r|$

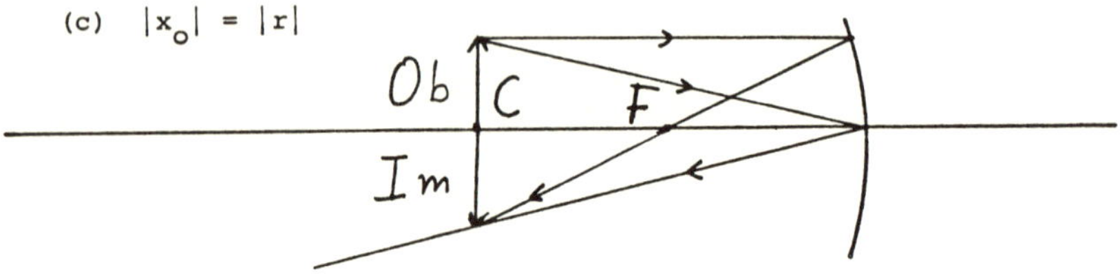

(d) $|r| \geq |x_o| > |f|$

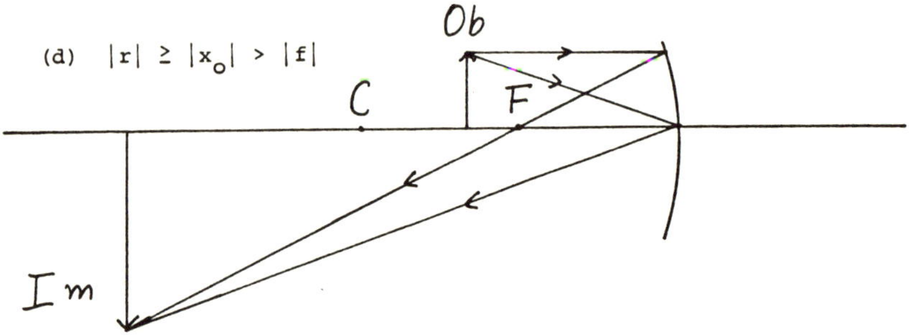

(e) $|x_o| < |f|$

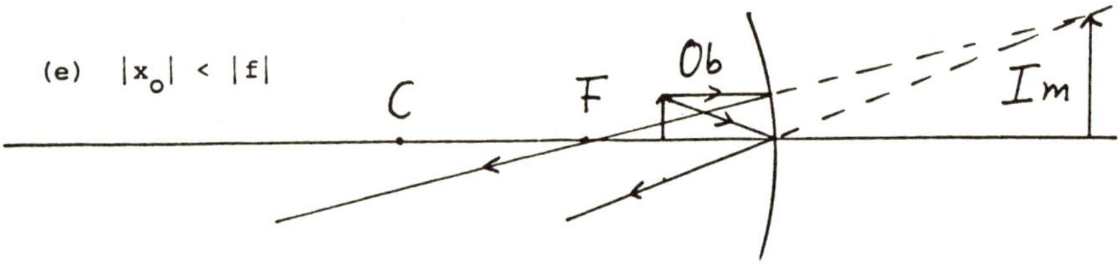

24. *Convex spherical mirror, analytical solution.* Use $r = 5$ cm and find x_i for the following:
   a. $x_o = -\infty$
   b. $|x_o| > r$, $x_o = -7$ cm
   c. $|x_o| = r$, $x_o = -5$ cm
   d. $r \geqq |x_o| > f$, $x_o = -4$ cm
   e. $|x_o| < f$, $x_o = -1$ cm

Solution

(a) $\underline{x_o = -\infty}$

$$\frac{1}{x_i} = \frac{1}{f} - \frac{1}{x_o} = \frac{1}{2.5} - \frac{1}{-\infty} \qquad \underline{x_i = 2.5\text{cm} = f}$$

(b) $\underline{|x_o| > r}$

$x_o = -7\text{cm}$

$$\frac{1}{x_i} = \frac{1}{2.5} - \frac{1}{-7} = \frac{1}{1.84} \qquad \underline{x_i = 1.84\text{cm}}$$

(c) $\underline{|x_o| = r}$

$$x_o = -5 \qquad \frac{1}{x_i} = \frac{1}{2.5} - \frac{1}{-5} = \frac{1}{1.67} \qquad \underline{x_i = 1.67\text{cm}}$$

(d) $\underline{r \geq |x_o| \geq f}$

$x_o = -4$ $\quad \frac{1}{x_i} = \frac{1}{2.5} - \frac{1}{-4} = \frac{1}{1.54}$ $\qquad \underline{x_i = 1.54\text{cm}}$

(e) $|x_o| < f$

$x_o = -1\text{cm}$ $\quad \frac{1}{x_i} = \frac{1}{2.5} - \frac{1}{-1} = \frac{1}{.714}$ $\qquad \underline{x_i = .714\text{cm}}$

25. *Convex spherical mirror, graphical solution.* Using the graphical method, find the image produced by a convex spherical mirror (r positive) for the following values of the object (to the left of the mirror):

a. $x_o = -\infty$
b. $|x_o| > r$
c. $|x_o| = r$
d. $r \geqq |x_o| > f$
e. $|x_o| < f$

Solution

Convex spherical mirror - graphical solution

(a) $x_o = -\infty$

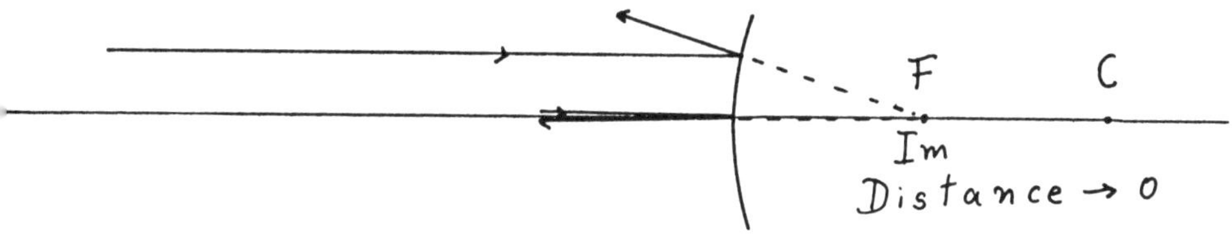

(b) $|x_o| > r$

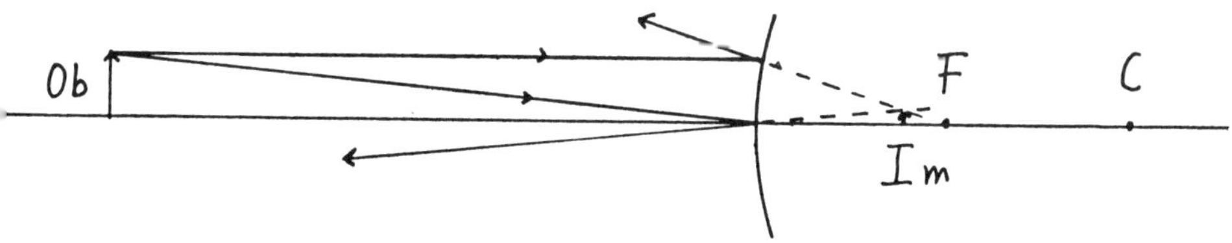

(c) $|x_o| = r$

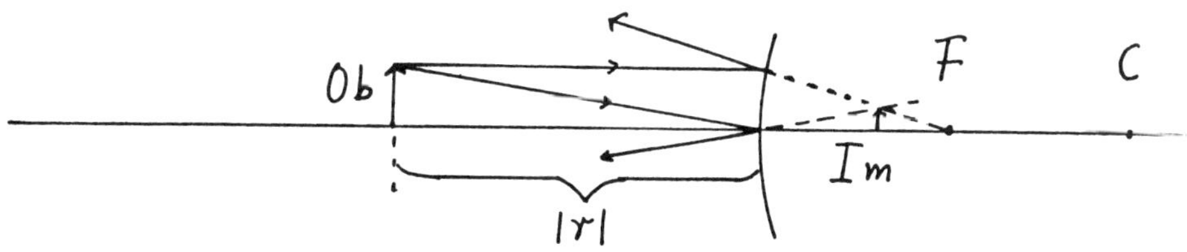

(d) $r \geq |x_o| > f$

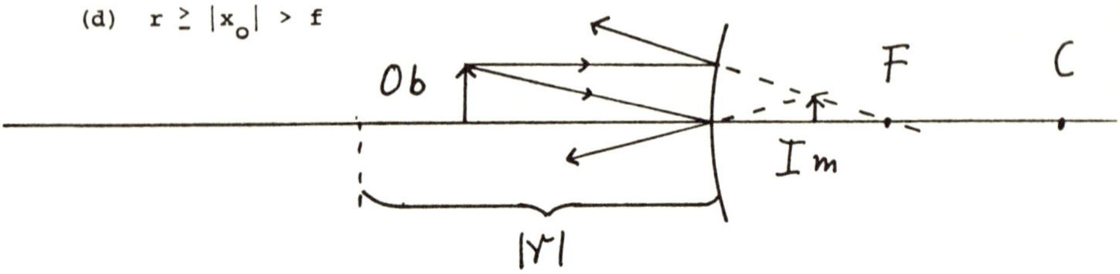

(e) $|x_o| < f$

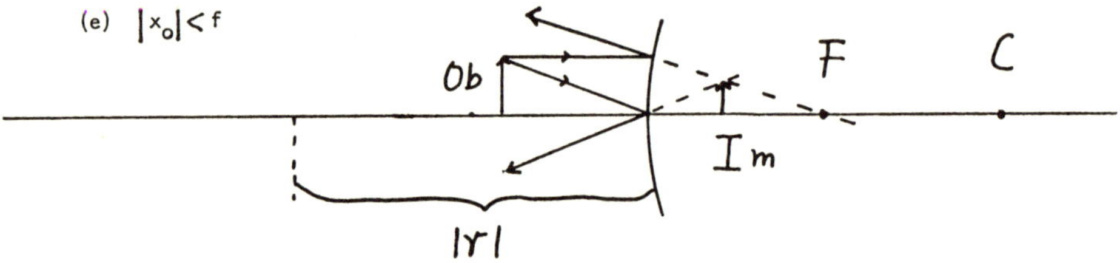

26. *Hemispherical cavity.* Consider the optical cavity consisting of a plane mirror (M_1) and a spherical mirror (M_2), separated by a distance d. The path of the ray, indicated in the figure shown here, retraces itself after eight steps. For clarity, the eight steps are shown in the figure as 1, 2, 3, …, 8.

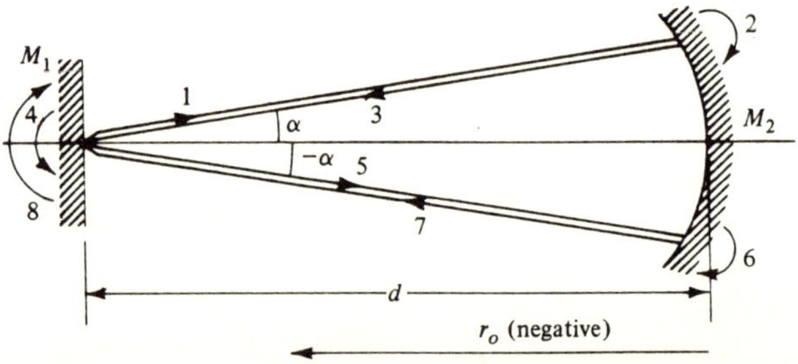

a. Write the eight matrices in the correct sequence and show that the resulting product is

$$M = \begin{pmatrix} \left(1 - \frac{2d}{r_o}\right)^2 - \frac{4d}{r_o}\left(1 - \frac{d}{r_o}\right) & 4d\left(1 - \frac{d}{r_o}\right)\left(1 - \frac{2d}{r_o}\right) \\ -\frac{4}{r_o}\left(1 - \frac{2d}{r_o}\right) & \left(1 - \frac{2d}{r_o}\right)^2 - \frac{4d}{r_o}\left(1 - \frac{d}{r_o}\right) \end{pmatrix}$$

b. The matrix is of the form $\begin{pmatrix} a & b \\ c & a \end{pmatrix}$; that is, we expect a double root from the determinant. Show that the eigenvalue of this problem may be expressed as $\lambda = a \pm \sqrt{bc}$ and give λ as a function of d/r.

c. For what value of d/r do we get $\lambda = 1$, and what does it mean?

Solution

Hemispherical cavity.

(a) Starting on the right

reflect. at 8	translation 6 → 8	reflect. at 6	translation 4 → 6	4, plane mirror reflect.	translation 2 → 4	refect. at 2	translation 1 → 2

$$\equiv \begin{pmatrix} 1 & 0 \\ 0 & -1 \end{pmatrix} \begin{pmatrix} 1 & -d \\ 0 & 1 \end{pmatrix} \begin{pmatrix} 1 & 0 \\ -\frac{2}{r} & -1 \end{pmatrix} \begin{pmatrix} 1 & d \\ 0 & 1 \end{pmatrix} \begin{pmatrix} 1 & 0 \\ 0 & -1 \end{pmatrix} \begin{pmatrix} 1 & -d \\ 0 & 1 \end{pmatrix} \begin{pmatrix} 1 & 0 \\ -\frac{2}{r} & -1 \end{pmatrix} \begin{pmatrix} 1 & d \\ 0 & 1 \end{pmatrix}$$

In successive pairs:

$$\begin{pmatrix} 1 & 0 \\ -\frac{2}{r} & -1 \end{pmatrix} \begin{pmatrix} 1 & d \\ 0 & 1 \end{pmatrix} = \begin{pmatrix} 1 & d \\ -2/r & -\frac{2d}{r} - 1 \end{pmatrix}$$

$$\begin{pmatrix} 1 & -d \\ 0 & 1 \end{pmatrix} \begin{pmatrix} 1 & d \\ -2/r & -\frac{2d}{r} - 1 \end{pmatrix} = \begin{pmatrix} 1+\frac{2d}{r} & d+\frac{2d^2}{r}+d \\ -2/r & -\frac{2d}{r} - 1 \end{pmatrix}$$

$$\begin{pmatrix} 1 & 0 \\ 0 & -1 \end{pmatrix} \begin{pmatrix} 1+\frac{2d}{r} & 2d+2d^2/r \\ -\frac{2}{r} & -\frac{2d}{r} - 1 \end{pmatrix} = \begin{pmatrix} 1+\frac{2d}{r} & 2d+\frac{2d^2}{r} \\ 2/r & \frac{2d}{r} + 1 \end{pmatrix}$$

$$\begin{pmatrix} 1 & d \\ 0 & 1 \end{pmatrix} \begin{pmatrix} 1+2d/r & 2d+2d^2/r \\ 2/r & \frac{2d}{r} + 1 \end{pmatrix} = \begin{pmatrix} 1+4d/r & 2d+\frac{2d^2}{r}+\frac{2d^2}{r}+d \\ 2/r & \frac{2d}{r} + 1 \end{pmatrix}$$

$$\begin{pmatrix} 1 & 0 \\ -\frac{2}{r} & -1 \end{pmatrix} \begin{pmatrix} 1+4d/r & 3d+\frac{4d^2}{r} \\ 2/r & \frac{2d}{r}+1 \end{pmatrix} = \begin{pmatrix} 1+\frac{4d}{r} & 3d+4d^2/r \\ -\frac{2}{r}-\frac{8d}{r^2}-\frac{2}{r} & -\frac{6d}{r}-\frac{8d^2}{r^2}-\frac{2d}{r}-1 \end{pmatrix}$$

$$\begin{pmatrix} 1 & -d \\ 0 & 1 \end{pmatrix} \begin{pmatrix} 1+\frac{4d}{r} & 3d+4d^2/r \\ -\frac{4}{r}-\frac{8d}{r^2} & -\frac{8d}{r}-\frac{8d^2}{r^2}-1 \end{pmatrix} =$$

$$= \begin{pmatrix} 1+\frac{4d}{r}+\frac{4d}{r} + \frac{8d^2}{r^2} & 3d+\frac{4d^2}{r}+\frac{8d^2}{r} + \frac{8d^3}{r^2}+d \\ -\frac{4}{r} - \frac{8d}{r^2} & -\frac{8d}{r}-\frac{8d^2}{r^2} - 1 \end{pmatrix}$$

$$\begin{pmatrix} 1 & 0 \\ 0 & -1 \end{pmatrix} \begin{pmatrix} 1+\frac{8d}{r}+\frac{8d^2}{r^2} & 4d+\frac{12d^2}{r}+\frac{8d^3}{r^2} \\ -\frac{4}{r} - \frac{8d}{r^2} & -\frac{8d}{r} - \frac{8d^2}{r^2} - 1 \end{pmatrix} =$$

$$\Rightarrow M = \begin{pmatrix} 1+\frac{8d}{r}+\frac{8d^2}{r^2} & 4d+\frac{12d^2}{r}+\frac{8d^3}{r^2} \\ \frac{4}{r}+\frac{8d}{r^2} & 1+\frac{8d}{r}+\frac{8d^2}{r^2} \end{pmatrix}$$

The second mirror is concave; so, $r \to -r_o$ with $r_o \geq 0$ i.e. the numerical radius will be given as a positive numerical No. Also $d \geq 0$.

$$\Rightarrow M = \begin{pmatrix} 1-\frac{8d}{r_o}+\frac{8d^2}{r_o^2} & 4d-\frac{12d^2}{r_o}+\frac{8d^3}{r_o^2} \\ \frac{-4}{r_o}+\frac{8d}{r_o^2} & 1-\frac{8d}{r_o}+\frac{8d^2}{r_o^2} \end{pmatrix}$$

factoring and completing squares:

$$M = \begin{pmatrix} 1-\frac{4d}{r_o}+\frac{4d^2}{r_o^2}-\frac{4d}{r_o}+\frac{4d^2}{r_o^2} & 4d[1-\frac{3d}{r_o}+\frac{2d^2}{r_o^2}] \\ \frac{-4}{r_o}(1-\frac{2d}{r_o}) & 1-\frac{4d}{r_o}+\frac{4d^2}{r_o}-\frac{4d}{r_o}+\frac{4d^2}{r_o^2}) \end{pmatrix}$$

$$\Rightarrow M = \begin{pmatrix} (1-\frac{2d}{r_o})^2-\frac{4d}{r_o}(1-\frac{d}{r_o}) & 4d(1-\frac{2d}{r_o})(1-\frac{d}{r_o}) \\ \frac{-4}{r_o}(1-\frac{2d}{r_o}) & (1-\frac{2d}{r_o})^2-\frac{4d}{r_o}(1-\frac{d}{r_o}) \end{pmatrix}$$

(b)

$$\begin{pmatrix} \ell_2 \\ \alpha_2 \end{pmatrix} = \overset{M}{\begin{pmatrix} m_{11} & m_{12} \\ m_{21} & m_{22} \end{pmatrix}} \begin{pmatrix} \ell_1 \\ \alpha_1 \end{pmatrix} = \begin{pmatrix} \lambda & 0 \\ 0 & \lambda \end{pmatrix} \begin{pmatrix} \ell_1 \\ \alpha_1 \end{pmatrix}$$

$$\Rightarrow \begin{pmatrix} m_{11}\ell_1 + m_{12}\alpha_1 \\ m_{21}\ell_1 + m_{22}\alpha_1 \end{pmatrix} = \begin{pmatrix} \lambda\ell_1 \\ \lambda\alpha_1 \end{pmatrix} \Rightarrow \begin{cases} m_{11}\ell_1 + m_{12}\alpha_1 = \lambda\ell_1 \\ m_{21}\ell_1 + m_{22}\alpha_1 = \lambda\alpha_1 \end{cases}$$

$$\Rightarrow \begin{cases} (m_{11}-\lambda)\ell_1 + m_{12}\alpha_1 = 0 \\ m_{21}\ell_1 + (m_{22}-\lambda)\alpha_1 = 0 \end{cases} \Rightarrow \begin{vmatrix} m_{11}-\lambda & m_{12} \\ m_{21} & m_{22}-\lambda \end{vmatrix} \equiv 0$$

$$\Rightarrow \quad (m_{11}-\lambda)(m_{22}-\lambda) - m_{21}m_{12} = 0$$

multiplying

$$\Rightarrow \quad m_{11}m_{22} - m_{11}\lambda - m_{22}\lambda + \lambda^2 - m_{12}m_{21} = 0$$

rearranging

$$\Rightarrow \lambda^2 - (m_{11} + m_{22})\lambda + [m_{11}m_{22} - m_{12}m_{21}] = 0$$

quadratic solm.

$$\lambda = \frac{+(m_{11}+m_{22})}{2} \pm \frac{\sqrt{(m_{11}+m_{22})^2 - 4(m_{11}m_{22} - m_{12}m_{21})}}{2}$$

Look at **M** and note $m_{11} = m_{22}$

substituting:

$$\Rightarrow \lambda = \frac{2m_{11}}{2} \pm \frac{\sqrt{4m_{11}^2 - 4m_{11}^2 + 4m_{12}\, m_{21}}}{2}$$

cancelling:

$$\Rightarrow \lambda = m_{11} \pm \sqrt{m_{12}\, m_{21}} \iff \lambda = a \pm \sqrt{bc}$$

with: $a \equiv m_{11}$

$b = m_{12}$

$c = m_{21}$

$$\lambda = [(1-\frac{2d}{r_o})^2 - \frac{4d}{r_o}(1-\frac{d}{r_o})] \pm \sqrt{-4d(1-\frac{d}{r_o})(1-\frac{2d}{r_o})\frac{4}{r_o}(1-\frac{2d}{r_o})}$$

<u>or</u>

$$\lambda = [(1-\frac{2d}{r_o})^2 - \frac{4d}{r_o}(1-\frac{d}{r_o})] \pm 4\sqrt{-\frac{d}{r_o}(1-\frac{d}{r_o})(1-\frac{2d}{r_o})^2}$$

<u>or</u>

$$\lambda = [|1-\frac{2d}{r_o}|^2 + \frac{4d}{r_o}(\frac{d}{r_o} - 1)] \pm 4|1-\frac{2d}{r_o}|\sqrt{\frac{d}{r_o}(\frac{d}{r_o} - 1)}$$

<u>or</u>

$$\lambda = [|\frac{2d}{r_o} - 1|^2 + \frac{4d}{r_o}(\frac{d}{r_o} - 1)] \pm 4|\frac{2d}{r_o} - 1|\sqrt{\frac{d}{r_o}(\frac{d}{r_o} - 1)}$$

For λ to be real: $d/r_o \geq 1$.

(c) At $d/r_o = 1$, one has $\lambda = \{[|2 - 1|^2 + 0] \pm 0\} = 1$

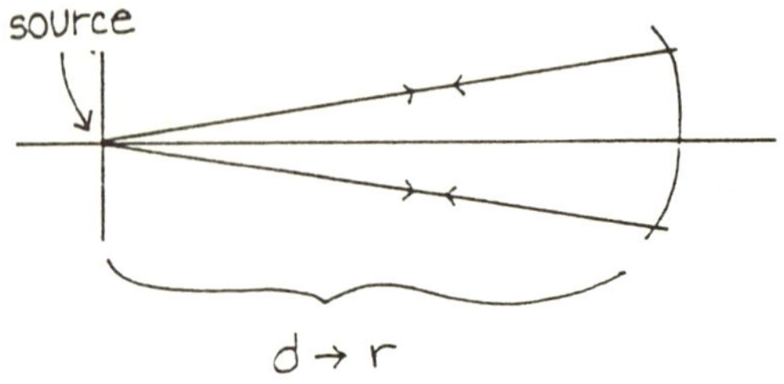

For a concave mirror a source at r will find its image back at the source. The rays will retroreflect off the concave mirror back to the same point on the plane mirror.

---

2 Interference of Light

1. *Time Average.* Calculate the time average in equation 2.12. The second factor in equation 2.12 is

$$a^2 = \cos^2\left[2\pi\left(\frac{x_1}{\lambda} - \frac{t}{T}\right) - 2\pi\frac{\delta/2}{\lambda}\right]$$

Make an appropriate choice for x_1 and show that we obtain

$$a_v^2 = \frac{1}{T}\int_0^T \cos^2(\ldots)\,dt = \frac{1}{2}$$

<u>Solution</u>

An appropriate choice for x_1 is $x_1 = \frac{\delta}{2}$ and we have for the time average

$$a^2 = \frac{1}{T}\int_0^T \cos^2\left(2\pi\frac{t}{T}\right) dt$$

we set $2\pi\frac{t}{T} = u$, $dt = \frac{T}{2\pi}\,du$ and have

$$a^2 = \frac{1}{2\pi}\int_0^{u=2\pi} \cos^2 u\,du = \frac{1}{2\pi}\left[\frac{u}{2} + \frac{1}{4}\sin 2u\right]_0^{2\pi}$$

$$= \frac{1}{2\pi}\left[\frac{2\pi}{2} + \frac{1}{4}\sin 4\pi - 0 - 0\right] = \frac{1}{2}$$

2. *Wedge-Shaped Film: Fringes of Equal Thickness.* Consider the fringes produced by a wedge-shaped film. This configuration can be produced by two microscope slides and a thin wire (diameter b). If we assume that the wedge angle α is small, then the bending of light rays can be neglected. The interference between light rays reflected from surfaces 2 and 3 results in a fringe pattern (see section 3.E.).

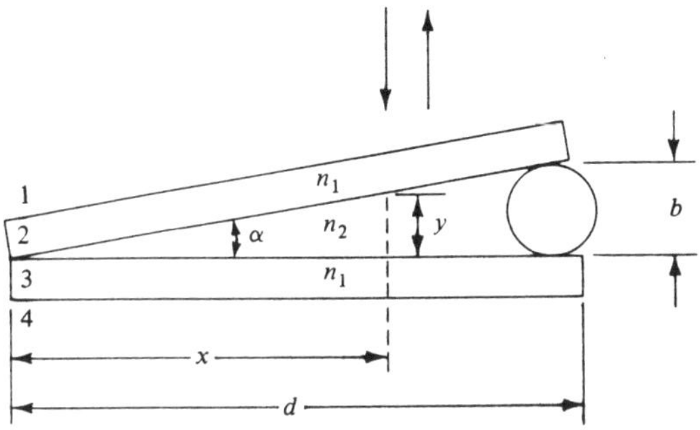

a. Use the derivation in the text and find the difference $\Delta x_m = x_{m+1} - x_m$ between successive maxima and the change $\Delta y_m = y_{m+1} - y_m$ in the film thickness from one maximum to the next.

b. Calculate the diameter b of the wire in micrometers if $\Delta x_m = 2$ mm, $\lambda = 0.5890\ \mu$m, and $d = 8$ cm.

Solution

Wedge-shaped film fringes, of equal thickness.

We want to get

(a) $\Delta x_m \equiv x_{m+1} - x_m$

From Eq. (2.26) for constructive interference

$$2x_m \tan\alpha + \frac{\lambda}{2} = m\lambda \Rightarrow x_m = (m\lambda - \frac{\lambda}{2}) \frac{1}{2\tan\alpha}$$

and

$$2x_{m+1} \tan\alpha + \frac{\lambda}{2} = (m+1)\lambda \Rightarrow x_{m+1} = (m\lambda + \frac{\lambda}{2}) \frac{1}{2\tan\alpha}$$

subtracting:

$$\Delta x_m \equiv x_{m+1} - x_m = \frac{\lambda}{2\tan\alpha}$$

For the change in film thickness from one maxima to the next:

From the figure: $\tan\alpha = \frac{y}{x}$

$$\Rightarrow y = x \tan\alpha \Rightarrow y_m = x_m \tan\alpha$$

$$y_{m+1} = x_{m+1} \tan\alpha$$

$\Rightarrow$ the change in thickness is (using Δ_m above):

$$(y_{m+1} - y_m) = (x_{m+1} - x_m) \tan\alpha = \frac{\lambda}{2}$$

(b) From the figure: $\tan\alpha = \frac{b}{d}$

From above: $\Delta x_m = \frac{\lambda}{2\tan\alpha}$

$$\Rightarrow b = \frac{\lambda d}{2(\Delta x_m)}$$

Numerically:

$$b = \frac{0.589\ (8 \times 10^4)}{2\ (2 \times 10^3)} = 12\mu m$$

3. *Newton's Rings.* The figure shows how fringes of equal thickness may be produced (Newton's rings). A spherical surface of index n_1 rests on a flat surface of the same index. A gap of variable thickness and index n_2 produces the interference fringes.

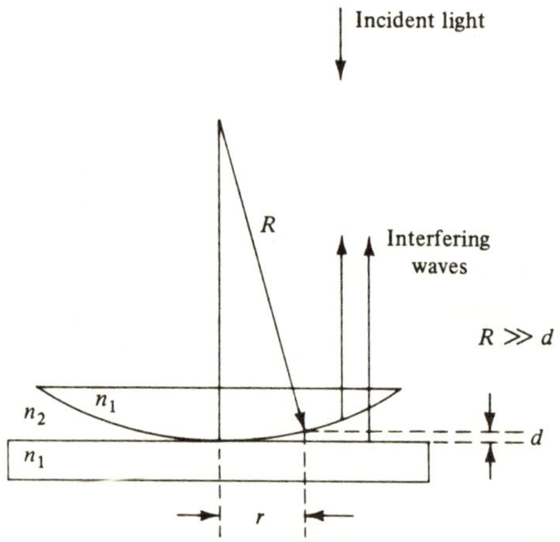

a. For constructive interference, give the dependence of d on m and λ.
b. Calculate r_m, the radius of the mth fringe as a function of m, λ, R, and n_2, and use the approximation $R \gg d$.
c. For $\lambda = 0.5\ \mu m$, $n_2 = 1$, and $R = 5$ m, calculate the radius of the first two rings.

Solution

Newton's Rings

(a) constructive interference (using the figure):

$$\left(2d_m + \frac{\lambda}{2n_2}\right) = \frac{m\lambda}{n_2} \qquad \left\{ \begin{array}{l} \text{in the medium } \lambda \to \frac{\lambda}{n_2} \\ m = 1,2,3,\ldots \end{array} \right.$$

↑ π phase shift on external reflection

solving for d:

$$d_m = \frac{1}{2}\left[\frac{\lambda}{n_2}\left(m - \frac{1}{2}\right)\right]$$

(b) Pythagoras => $R^2 = r^2 + (R-d)^2 = r^2 + R^2 + d^2 - 2Rd$

the R's cancel and $d^2 \ll R^2$: $\Rightarrow d \approx \frac{r^2}{2R} \Rightarrow d_m = \frac{r_m^2}{2R}$

Combining from part A:

$$\frac{r_m^2}{2R} = \frac{\lambda}{2n_2}\left(m-\frac{1}{2}\right) \Rightarrow r_m = \sqrt{\frac{\lambda R(m-\frac{1}{2})}{n_2}}$$

(c) Numerical:

$$r_{m=1} = \sqrt{\frac{0.5(5\text{x}10^{+6})1/2}{1}} = 1.12 \text{ x } 10^3 \mu\text{m} = 1.12\text{mm}$$

$$r_{m=2} = \sqrt{\frac{0.5(5\text{x}10^{6})(2-1/2)}{1}} = 1.94 \text{ x } 10^3 \mu\text{m} = 1.94\text{mm}$$

4. *Complex numbers.* We saw how the superposition of two waves in the cosine form may be accomplished with complex numbers to obtain the square of the amplitude. Here we summarize the amplitude factor in question.

$$y_1 = A\cos\alpha_1$$
$$y_2 = A\cos(\alpha_1 - \Delta)$$

For the sum $y_1 + y_2$, we get

$$y = y_1 + y_2 = 2A\cos\frac{\Delta}{2}\cos\left(\alpha_1 - \frac{\Delta}{2}\right)$$

Using complex numbers

$$z_1 = Ae^{i\alpha_1}$$
$$z_2 = Ae^{i(\alpha_1 - \Delta)}$$

we have

$$z = z_1 + z_2 = 2A\cos\frac{\Delta}{2}e^{i(\alpha_1 - \Delta/2)}$$

Now let us consider waves in the sine form.

$$y_1 = A\sin\alpha_1, \quad y_2 = A\sin(\alpha_1 - \Delta)$$

a. Show that by adding y_1 and y_2 and using

$$\sin\gamma + \sin\beta = 2\sin\frac{\gamma+\beta}{2}\cos\frac{\gamma-\beta}{2}$$

the same amplitude factor is obtained. Underline it.

b. Use complex numbers. By complementing y_1 and y_2 to become complex numbers

$$z_1 = Ae^{i\alpha_1} \quad \text{and} \quad z_2 = Ae^{i(\alpha_1 - \Delta)}$$

we obtain the result given above,

$$z = z_1 + z_2 = 2A\cos\frac{\Delta}{2}e^{(\alpha_1 - \Delta/2)}$$

Show that the result of (a) is the imaginary part of z in (b).

Solution

1. Superposition of two waves, sine functions

(a) Trigonometry approach:

$$y = y_1 + y_2 = A\{\sin\alpha_1 + \sin(\alpha_1 - \Delta)\}$$

$$= 2A\sin\left[\frac{\alpha_1 + (\alpha_1 - \Delta)}{2}\right]\cos\left[\frac{\alpha_1 - (\alpha_1 - \Delta)}{2}\right]$$

$$= 2A\sin\left(\alpha_1 - \frac{\Delta}{2}\right)\cos\left(\frac{\Delta}{2}\right)$$

$$= \underline{2A\cos\left(\frac{\Delta}{2}\right)}\sin\left(\alpha_1 - \frac{\Delta}{2}\right)$$

(b) Complex No. approach:

$$z = z_1 + z_2 = Ae^{i\alpha_1} + Ae^{i(\alpha_1 - \Delta)}$$

$$= 2A\cos\left(\frac{\Delta}{2}\right)e^{i(\alpha_1 - \frac{\Delta}{2})}$$

$$y = y_1 + y_2 = \text{Im } z = \text{Im } z_1 + \text{Im } z_2$$

$$= \text{Im } 2A\cos\left(\frac{\Delta}{2}\right)e^{i(\alpha_1 - \frac{\Delta}{2})}$$

$$= \underline{2A\cos\left(\frac{\Delta}{2}\right)}\text{ Im } e^{i(\alpha_1 - \frac{\Delta}{2})}$$

This is the same result $= \underline{2A\cos(\frac{\Delta}{2})}\sin(\alpha_1 - \frac{\Delta}{2})$ as found by the trigonometry approach above.

5. *General Case: Different Amplitudes, Different Phases, Sine Notation.* We want to calculate the superposition of the following two functions having different amplitudes and different phases:

$$u_1 = u_{10}\sin(\omega t - kx_1 + \delta_1)$$

and

$$u_2 = u_{20}\sin(\omega t - kx_2 + \delta_2)$$

If we attempt to evaluate $u^2 = (u_1 + u_2)^2$ directly, we encounter some complications. To avoid this, let $\alpha_i = -kx_i + \delta_i$, $i = 1, 2$, and calculate

$$u = u_1 + u_2 = u_{10}\sin(\omega t + \alpha_1) + u_{20}\sin(\omega t + \alpha_2) = u_0\sin(\omega t + \alpha)$$

where

$$u_0\cos\alpha = u_{10}\cos\alpha_1 + u_{20}\cos\alpha_2$$

and

$$u_0\sin\alpha = u_{10}\sin\alpha_1 + u_{20}\sin\alpha_2$$

Since

$$(u_0\cos\alpha)^2 + (u_0\sin\alpha)^2 = u_0^2$$

we get

$$u_0^2 = u_{10}^2 + u_{20}^2 + 2u_{10}u_{20}\cos(\alpha_1 - \alpha_2)$$

and therefore

$$u = [u_{10}^2 + u_{20}^2 + 2u_{10}u_{20}\cos(\alpha_1 - \alpha_2)]^{1/2}\sin(\omega t + \alpha)$$

a. Fill in all the missing steps in the calculations leading to the final equation.
b. Express $\alpha_1 - \alpha_2$ in terms of k, x_i and δ_i, where $i = 1, 2$.
c. The conditions for interference maxima and minima are determined by the amplitude factor in the final equation for u, that is, the square root term. Find these conditions of u and obtain a general expression involving k, x_i, and δ_i for these two cases.
d. For the special case of $u_{10} = u_{20}$ (i.e., equal amplitude waves), show that the expression for u leads to the same result for the intensity u^2 as was obtained in equation 2.36 using complex notation.

Solution

General case; different amplitudes, different phases, sine notation.

(a) Two waves meeting the criteria

$$\begin{cases} u_1 = u_{1o} \sin(\omega t - kx_1 + \delta_1) \\ u_2 = u_{2o} \sin(\omega t - kx_2 + \delta_2) \end{cases}$$

let: $\begin{Bmatrix} \alpha_1 = -kx_1 + \delta_1 \\ \alpha_2 = -kx_2 + \delta_2 \end{Bmatrix} \Rightarrow \begin{Bmatrix} u_1 = u_{1o} \sin(\omega t + \alpha_1) \\ u_2 = u_{2o} \sin(\omega t + \alpha_2) \end{Bmatrix}$

$$u \equiv u_1 + u_2 = u_{1o} \sin(\omega t + \alpha_1) + u_{2o} \sin(\omega t + \alpha_2)$$

$$= \{u_{1o}[\sin \omega t \cos \alpha_1 + \cos \omega t \sin \alpha_1]$$

$$+ u_{2o}[\sin \omega t \cos \alpha_2 + \cos \omega t \sin \alpha_2]\}$$

factoring $\sin \omega t$ and $\cos \omega t$ terms:

$$u = \{(u_{1o} \cos \alpha_1 + u_{2o} \cos \alpha_2) \sin \omega t$$

$$+ (u_{1o} \sin \alpha_1 + u_{2o} \sin \alpha_2) \cos \omega t\}$$

defining u_o, $\cos \alpha$, $\sin \alpha$:

$$= (u_o \cos \alpha) \sin \omega t + (u_o \sin \alpha) \cos \omega t$$

factoring & trigonometry:

$$u = u_o \{\cos \alpha \sin \omega t + \sin \alpha \cos \omega t\}$$

$$= u_o \sin(\omega t + \alpha)$$

For the intensity: $u^2 = u_o^2 \sin^2(\omega t + \alpha)$

Now what is u_o^2 under these definitions? From above using our definitions of u_o and α:

$$\begin{cases} u_o \cos\alpha = u_{1o}\cos\alpha_1 + u_{2o}\cos\alpha_2 \\ u_o \sin\alpha = u_{1o}\sin\alpha_1 + u_{2o}\sin\alpha_2 \end{cases}$$

$$\Rightarrow u_o^2 \cos^2\alpha + u_o^2 \sin^2\alpha = u_o^2$$

substi. $u_o^2 = \{[u_{1o}^2 \cos^2\alpha_1 + u_{2o}^2 \cos^2\alpha_2 + 2u_{1o}u_{2o}\cos\alpha_1\cos\alpha_2]$

$$+ [u_{1o}^2 \sin^2\alpha_1 + u_{2o}^2 \sin\alpha_2 + 2u_{1o}u_{2o}\sin\alpha_1\sin\alpha_2]\}$$

collecting like u_{1o}^2, u_{2o}^2, and $u_{1o}u_{2o}$ terms:

$$u_o^2 = \{(\cos^2\alpha_1 + \sin^2\alpha_1)u_{1o}^2 + (\cos^2\alpha_2 + \sin^2\alpha_2)u_{2o}^2$$

$$+ 2u_{1o}u_{2o}(\cos\alpha_1\cos\alpha_2 + \sin\alpha_1\sin\alpha_2)\}$$

Substituting u_o^2 in the Eq. for the intensity:

$$u^2 = \{u_{1o}^2 + u_{2o}^2 + 2u_{1o}u_{2o}\cos(\alpha_1-\alpha_2)\}\sin^2(\omega t + \alpha).$$

$$u = \{u_{1o}^2 + u_{2o}^2 + 2u_{1o}u_{2o}\cos(\alpha_1-\alpha_2)\}^{1/2}\sin(\omega t + \alpha)$$

(b) From the beginning of part (a)

$$(\alpha_1 - \alpha_2) = -kx_1+\delta_1+kx_2-\delta_2 = [-k(x_1-x_2)+(\delta_1-\delta_2)]$$

(c) $\sin(\omega t+\alpha)$ is the traveling wave part with α = (phase) of the total wave.

The maxima and minima are found in the term:

$$\{u_{1o}^2 + u_{2o}^2 + 2u_{1o}u_{2o}\cos(\alpha_1-\alpha_2)\}^{1/2};\ u_{1o} \text{ and } u_{2o} \geq 0$$

the maxima $\Rightarrow \cos(\alpha_1-\alpha_2) = +1 \Rightarrow (\alpha_1-\alpha_2) = m2\pi$

where $m = 0,1,2,\ldots$ or $(\alpha_1-\alpha_2) = -k(x_1-x_2) + (\delta_1-\delta_2) = m2\pi$

the minima $\Rightarrow \cos(\alpha_1-\alpha_2) = -1 \Rightarrow (\alpha_1-\alpha_2) = (2m+1)\pi$

where $-m = 0,-1,-2,\ldots$ or $(\alpha_1-\alpha_2) = -k(X_1-X_2) + (\delta_1-\delta_2)$
$= (2m + 1)\pi$

(d) Assume $u_{1o} = u_{2o} \equiv A$

from the end of part (A) we obtain

$$u = \sqrt{2}A\{1 + \cos(\alpha_1 - \alpha_2)\}^{1/2}\sin(\omega t + \alpha)$$

Trigonometry $\Rightarrow 1 + \cos q = 2\cos^2(q/2)$

$$\Rightarrow u = 2A\cos\left(\frac{\alpha_1-\alpha_2}{2}\right)\sin(\omega t+\alpha)$$

We use $(\alpha_1 - \alpha_2) \equiv \Delta$

Substituting

$$u = 2A\cos\left(\frac{\Delta}{2}\right)\sin(\omega t+\alpha)$$

Dropping the time factor, we have

$$u^2 = 4A^2\cos^2\left(\frac{\Delta}{2}\right)$$

6. *N-Source Interference Pattern.* Consider the case $N = 5$. Make a sketch of the intensity as a function of Y/X. Plot the center and the first two main maxima along with all secondary maxima in between.

Part. ans. Major peaks at $Y/X = (\lambda/b)m$, zeros at $Y/X = m\lambda/5b$ (except at major peaks)

Solution

N - source interference pattern.

We have for N = 5

$$\frac{u^*u}{A^2} = \frac{\sin^2\left(\pi 5\frac{b}{\lambda}\frac{Y}{X}\right)}{\sin^2\left(\pi\frac{b}{\lambda}\frac{Y}{X}\right)}$$ where b is the distance between the sources.

The numerator and denominator equal zero when

$$\pi\frac{b}{\lambda}\frac{Y}{X} = 0, \pi, 2\pi, \ldots m\pi \quad \text{where:} \quad m = 0, 1, 2, \ldots$$

The indeterminate fraction 0/0 appears; with l'Hopital's rule yielding $(u^*u/A^2 = N^2 = 25$.

(a) This means the major peaks are at $(Y/X) = \frac{\lambda}{b} m$.

(b) The minor peaks are near $(5\pi\frac{b}{\lambda}\frac{Y}{X}) = m\frac{\pi}{2} \Rightarrow \frac{Y}{X} = \frac{m}{10}\frac{\lambda}{b}$

except at the zeros (c) and peaks (a)

(c) The zeros are at $(5\pi\frac{b}{\lambda}\frac{Y}{X}) = m\pi \Rightarrow \frac{Y}{X} = \frac{m\lambda}{5b}$ except at the major peaks of (a)

Note: The minor peaks immediately adjacent to the major peaks, on each side, are indistinguishable.

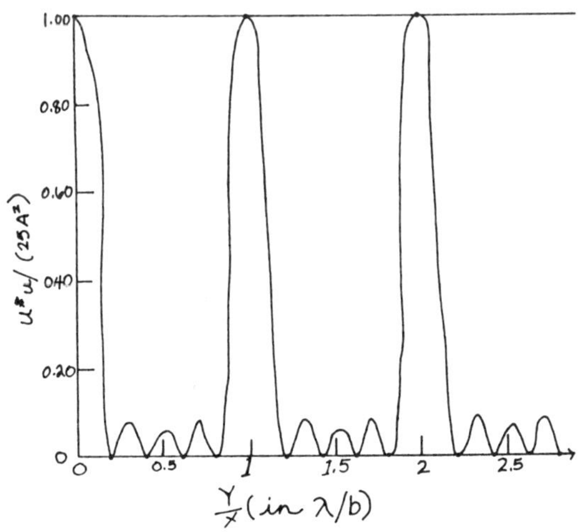

7. *Plane Parallel plate. Three Media.* A plane parallel plate of thickness D and refraction index n_2 is surrounded by two media with indices n_1 and n_3. Consider only two waves reflected from the interfaces, as indicated in the following figure:

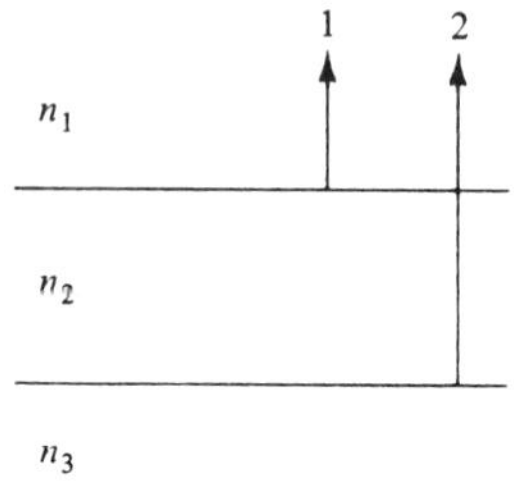

Calculate the path difference and phase difference of waves (1) and (2). For example, for an air gap surrounded by glass, where $n_1 = 1.5$, $n_2 = 1$, and $n_3 = 1.5$, one has

$$\delta = 2D + \frac{\lambda}{2} \quad \text{and} \quad \Delta = \frac{2\pi}{\lambda}\left(2D + \frac{\lambda}{2}\right) = 2\pi\left(\frac{2D}{\lambda}\right) + \pi$$

a. For a glass plate in air: $n_1 = 1$, $n_2 = 1.5$, and $n_3 = 1$
b. For three media with increasing refractive indices, so that $n_2^2 = n_1 \cdot n_3$, with $n_1 = 1$, $n_2 = 1.22$, and $n_3 = 1.5$ (antireflection coating if D is $\lambda/4$).
c. For the case where $n_1 = 1$, $n_2 = 1.5$, and $n_3 = 1.22$.
d. For the case where $n_1 = 1.5$, $n_2 = 1$, and $n_3 = 1.22$.
e. For the case where $n_1 = 1.5$, $n_2 = 1.22$, and $n_3 = 1$.

Solution

(a) $\delta = 2D + \frac{\lambda}{2}$ $\qquad \Delta = 2\pi\left(\frac{2D}{\lambda}\right) + \pi$ $\qquad \frac{2\pi}{\lambda/n} 2D + \pi$

(b) $\delta = 2D$ $\qquad \Delta = 2\pi\left(\frac{2D}{\lambda}\right)$

(c) $\delta = 2D + \frac{\lambda}{2}$ $\qquad \Delta = 2\pi\left(\frac{2D}{\lambda}\right) + \pi$

(d) $\delta = 2D + \frac{\lambda}{2}$ $\qquad \Delta = 2\pi\left(\frac{2D}{\lambda}\right) + \pi$

(e) $\delta = 2D$ $\qquad \Delta = 2\pi\left(\frac{2D}{\lambda}\right)$

8. *Michelson Interferometer Circular Fringes.* The interference of the beams in a Michelson interferometer was discussed for two beams incident and reflected under 45° to the normal of the beam splitter. We will call these two beams the axial beams. For the intensity, we obtained (equation 2.21),

$$y^2 = 4A^2 \cos^2\left(2\pi \frac{\delta/2}{\lambda}\right)$$

and the conditions for maxima were

$$\delta = 2(x_2 - x_1) = m\lambda$$

and for minima

$$\delta = 2(x_2 - x_1) = \left(m + \frac{1}{2}\right)\lambda$$

Now we want to discuss the interference of beams with slightly larger or smaller angles than 45° leading to a ring pattern.

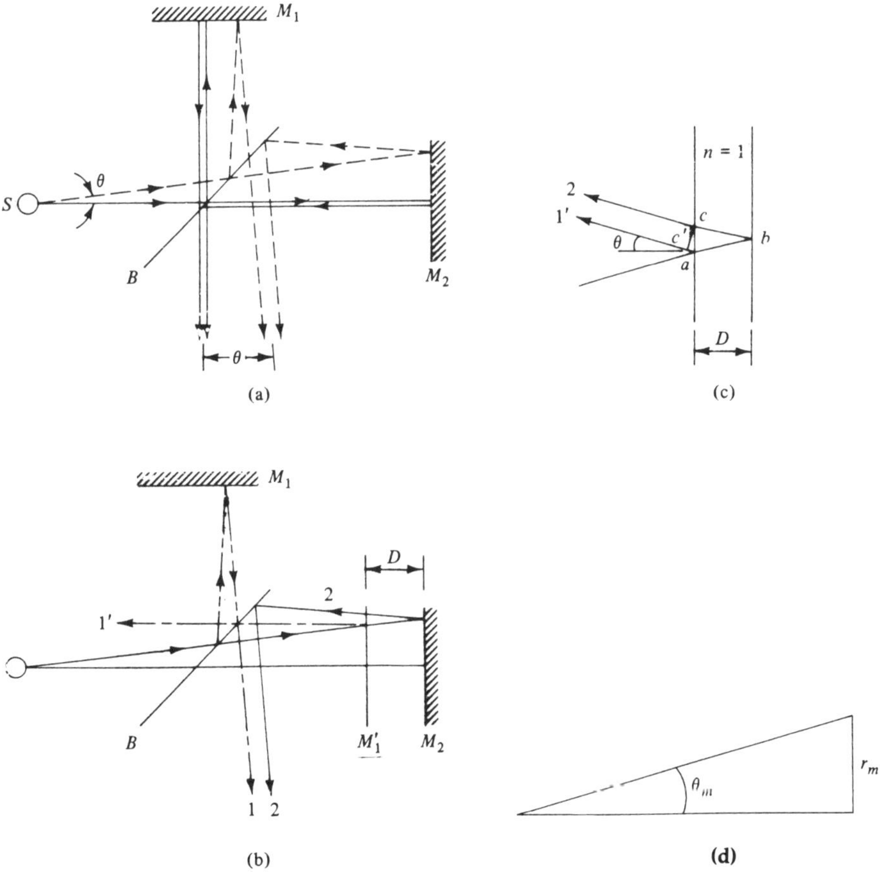

Part (a) of the accompanying figure, shows two beams leaving the source at different angles. For the axial beams, the superposition produces the usual result. For the beams leaving the source with angle θ, the resulting superposition of the beams has the angle θ with the axial beams (in direction of the observation screen). To obtain the conditions for maxima and minima for the superposition of the nonaxial beam, we look at part (b). Only the superposition of two nonaxial beams are shown. The plane M_1' indicates the distance beam 1 travels (with respect to M_2) compared to beam 2. If we "fold" the path of beam 1 around the beam splitter, we see that M_1 and M_1' are at the same position. Therefore, beams 1 and 1′ are equivalent. The difference in optical path is shown in figure (c) and is the same as obtained for the plane parallel plate (see equation 2.58),

$$\delta = 2D\cos\theta$$

where $n = 1$.

a. Find the conditions for maxima and minima. Beams emitted with the same angle θ from the source form a cone. The superposition that results also forms a cone in the direction of the observation screen. Since the optical path difference depends on θ, we observe a ring pattern at the screen. At a plane parallel plate, the corresponding fringes are called fringes of equal inclination. The same term is also used here.

b. Calculate all allowed values of θ of maxima for the following cases:
   (i) $D/\lambda = 1$
   (ii) $D/\lambda = 2$
   (iii) $D/\lambda = 4$

c. Make a sketch of the ring pattern. The radius is proportional to $\sin\theta$.
   Note: The fringe with the highest m is closest to the axis. This is in contrast to the interference pattern of N point sources.

Solution

(a) $y^2 = 4A^2 \cos^2 (\pi\frac{\delta}{\lambda}) = 4A^2 \cos^2 (2\pi \frac{D\cos\theta}{\lambda})$

Maxima $\frac{2\pi D\cos\theta}{\lambda} = m\pi$ or $m\lambda = 2D\cos\theta$

Minima $\frac{2\pi D\cos\theta}{\lambda} = (1,3,5\ldots)\frac{\pi}{2}$ or $(m+\frac{1}{2})\lambda = 2D\cos\theta$

$m = 0, 1, 2, 3$

(b) (i) $\frac{D}{\lambda} = 1$

Maxima: $2\pi\cos\theta = m\pi$

$\cos\theta = \frac{m}{2}$

allowed: $m = 0$, $m = 1$, $m = 2$

corresponding to $\theta = 90°, 60°, 0°$

(ii) $\frac{D}{\lambda} = 2$

Maxima: $4\pi\cos\theta = m\pi$

$\cos\theta = \frac{m}{4}$

allowed: $m = 0,1,2,3,4$

corresponding to $\theta = 90°, 75.5°, 60°, 41.4°, 0°$

(iii) $\frac{D}{\lambda} = 4$

Maxima: $8\pi\cos\theta = m\pi$

$\cos\theta = \frac{m}{8}$

allowed: $m = 0,1,2,3,4,5,6,7,8$

corresponding to $\theta = 90°$, $82.8°$, $75.5°$, $68°$, $60°$, $51.3°$, $41.4°$, $29°$, $0°$

(c)

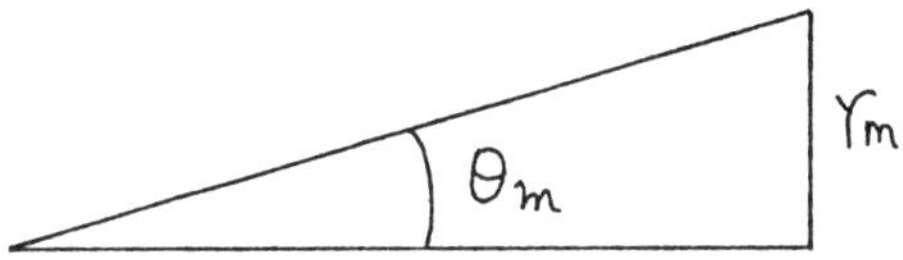

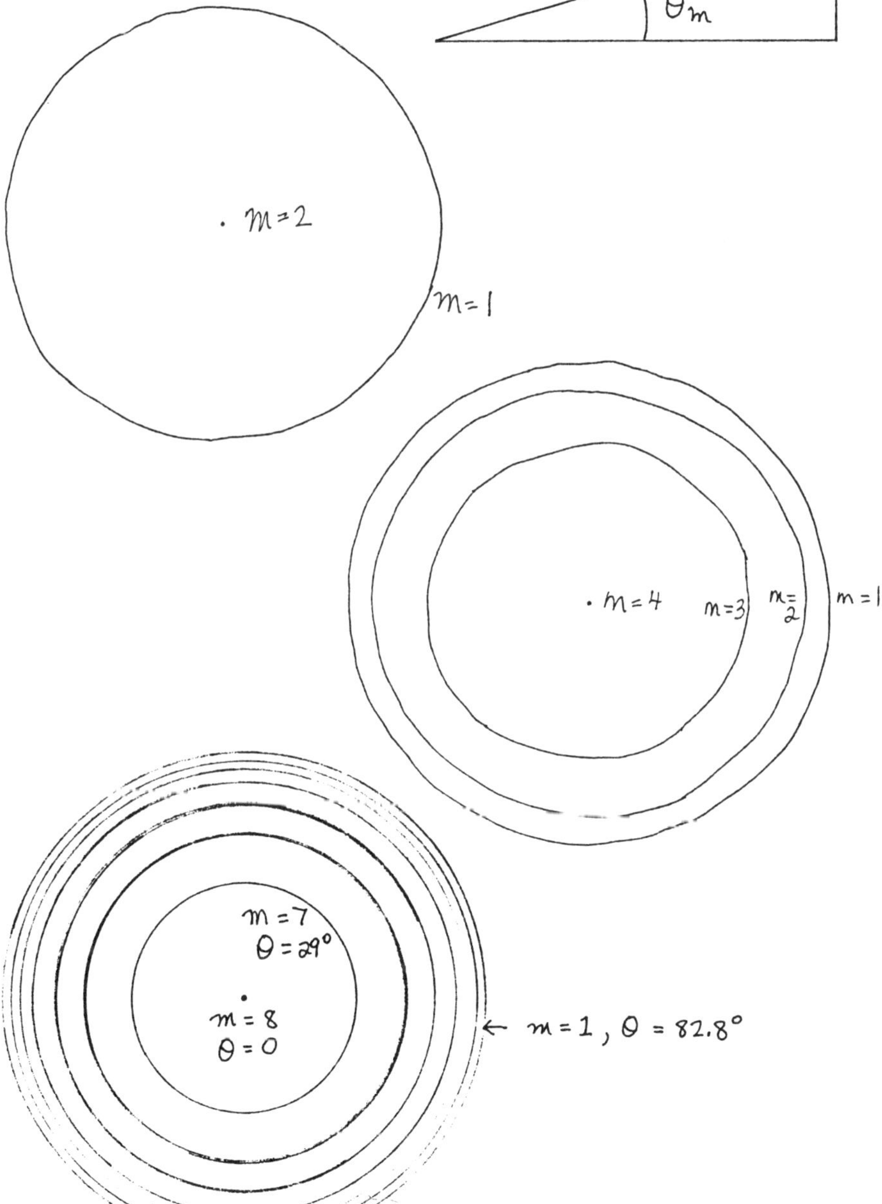

9. *Plane Parallel Plate: Multiple-Beam Interference.* For the transmitted intensity equation 2.69),

$$I_t = \frac{1}{1 + g^2 \sin^2 \Delta/2}$$

where g is defined as $g = 2r/(1 - r^2)$.

This formula is also valid for the case of two highly reflecting plates; that is, r may have values close to 1. For large values of r, the device is called a Fabry-Perot etalon. The quantity Δ is defined as (equation 2.71)

$$\Delta = \frac{2\pi}{\lambda} 2nD \cos\theta_2$$

We would like to investigate I_t as a function of θ, holding λ and r constant. This will result in fringes of constant inclination. Consider light incident on a Fabry-Perot etalon at an angle θ with respect to the normal (see the figure). The two surfaces of the Fabry-Perot etalon have constant reflectivity r and are separated by a distance D. The index of refraction between the two surfaces is assumed to be a little larger than 1 (to remain a denser medium); we take $\theta = \theta_2$.

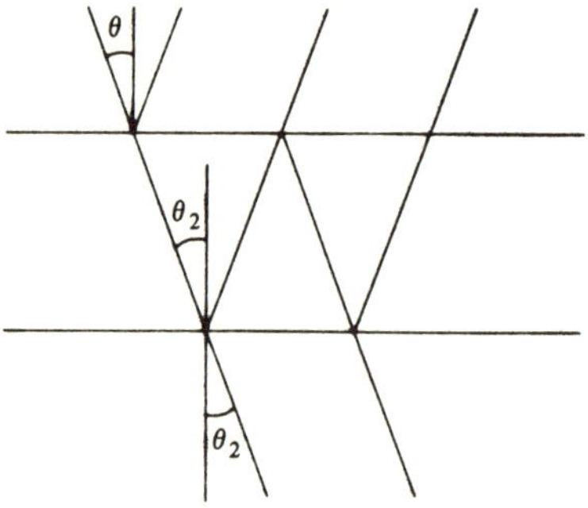

a. Determine the angle θ at which constructive interference is observed on a distant screen, that is, when $I_t = 1$. Find all possible angles (of fringes) for the following cases:

   (i) $D/\lambda = 1$ (ii) $D/\lambda = 2$ (iii) $D/\lambda = 4$

b. Plot m versus θ for case (iii). Note that the smallest number of m corresponds to the largest angle.

Solution

(a) $I_t = 1 \rightarrow \sin^2 \frac{\Delta}{2} = 0 \rightarrow \frac{\Delta}{2} = m\pi$

$\rightarrow \frac{\pi}{\lambda} 2D\cos\theta_2 = m\pi \rightarrow 2D \cos\theta_2 = m\lambda$

m integer

(i) $\frac{D}{\lambda} = 1$ $\cos\theta_2 = \frac{m}{2} \rightarrow m = 0,1,2$

$\rightarrow \theta_2 = 90°, 60°, 0°$

(ii) $\frac{D}{\lambda} = 2$ $\cos\theta_2 = \frac{m}{4}$ $\rightarrow$ $m = 0,1,2,3,4$

$\theta_2 = 90°,\ 75.5°,\ 60°,\ 41.4°, 0°$

(iii) $\frac{D}{\lambda} = 3$ $\cos\theta_2 = \frac{m}{8}$ $\rightarrow m = 0,1,2,3,4,5,6,7,8$

$\theta_2 = 90°,\ 82,8°,\ 75.5°,\ 68°,\ 60°$

$51.3°,\ 41.4°,\ 29°,\ 0°$

(b)

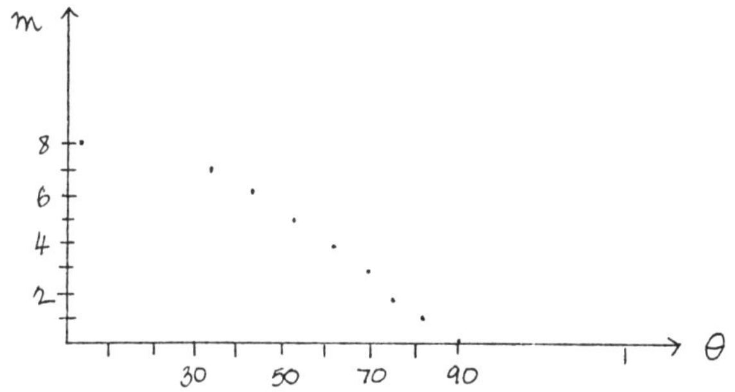

10. *Plane Parallel Plate: Width of Half Height.* For the plane parallel plate, transmitted intensity is given as

$$I_t = \frac{1}{1 + g^2 \sin^2 \Delta/2}$$

where

$$g^2 = \frac{4r^2}{(1 - r^2)^2}$$

and r is the reflection coefficient of the amplitude at one surface. For Δ we have

$$\frac{\Lambda}{2} = \frac{2\pi}{\lambda} nD \cos\theta_2$$

where n is the refractive index, D is the thickness of the plate, λ is the wavelength, and θ_2 is the angle with respect to the normal in the plate.

a. Make a sketch of I_t versus λ for $\lambda = 2$ mm to $\lambda = \frac{1}{4}$ mm. Set $n = 1$, and $\theta_2 = 0$, and use $D = 1$ mm. Compare with Figure 2.24, where we plotted I_t versus $\Delta/2 \propto D/\lambda$. Note that in Figure 2.24, the width at half height of the lines depends on r.
b. Show that we have for the width $\Delta\lambda$ at half height of the lines

$$\frac{\lambda}{\Delta\lambda} = \frac{m\pi g}{2} = \frac{m\pi r}{1 - r^2}$$

(The calculation is similar to that shown in Chapter 8 for the resolution of a Fabry-Perot spectrometer)

Solution

Plane parallel plate. Width at half height.

(a)

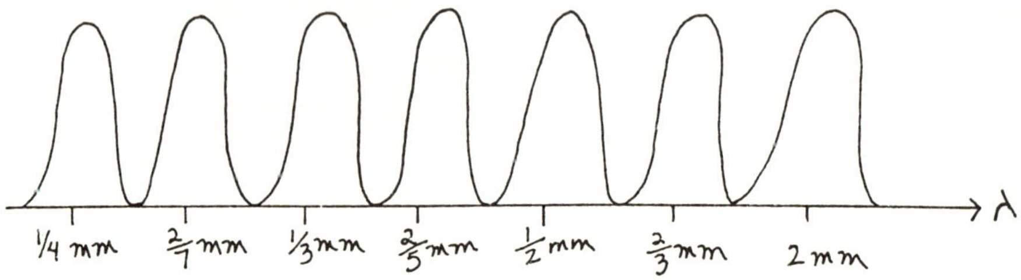

Max for $\sin^2 \frac{\Delta}{2} = 0$

that is $\frac{\Delta}{2} = 0, \pi, 2\pi$ ·

that is $\frac{2\pi(1\text{ mm})}{\lambda} = 0, \pi, 2\pi, 3\pi, 4\pi, 5\pi$

or $\frac{1\text{ mm}}{\lambda} = 0, \frac{1}{2}, 1, \frac{3}{2}, 2, \frac{5}{2}$ etc.

$1\text{ mm} = \lambda(0, \frac{1}{2}, 1, \frac{3}{2}, 2, \frac{5}{2})$

$$\lambda = \frac{1\text{ mm}}{\left(0, \frac{1}{2}, 1, \frac{3}{2}, 2, \frac{5}{2}, 3, \frac{7}{2}, 4\right)}$$

$\lambda = \infty$

$\lambda = 2\text{mm}$

$\lambda = 1\text{mm}$

$\lambda = \frac{2\text{mm}}{3}$

$\lambda = \frac{1}{2}\text{ mm}$

$\lambda = \frac{2}{5}\text{ mm}$

$\lambda = \frac{1}{3}\text{ mm}$

$\lambda = \frac{2}{7}\text{ mm}$

$\lambda = \frac{1}{4}\text{ mm}$

(b)

$$\frac{1}{1 + g^2 \sin^2 \frac{\Delta}{2}} = \frac{1}{2} \quad / \quad \frac{\Delta}{2} = \frac{2\pi D}{\lambda} \qquad (1)$$

We call the phase difference from the maximum to the half height point on the right and left $\pm\varepsilon$ and have

$$\frac{2\pi}{\lambda_1} D = \frac{m\pi}{2} + \varepsilon \, , \quad \frac{2\pi}{\lambda_2} D = \frac{m\pi}{2} - \varepsilon$$

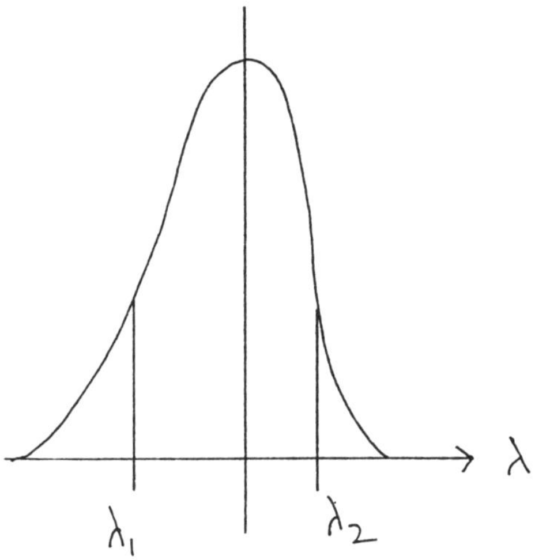

We subtract $\quad \frac{2\pi D}{\lambda_2} - \frac{2\pi D}{\lambda_1} = - 2\,\varepsilon$

$$\frac{1}{\lambda_2} - \frac{1}{\lambda_2 - \Delta\lambda} = - \frac{\varepsilon}{\pi D}$$

$$\frac{1}{\lambda_2 - \Delta\lambda} = \frac{1}{\lambda_2} \left(\frac{1}{1 - \frac{\Delta\lambda}{\lambda_2}}\right) \simeq \frac{1}{\lambda_2} \left(1 + \frac{\Delta\lambda}{\lambda_2}\right) = \frac{1}{\lambda_2} + \frac{\Delta\lambda}{\lambda_2^{\,2}}$$

We get

$$\frac{1}{\lambda_2} - \frac{1}{\lambda_2} - \frac{\Delta\lambda}{\lambda_2^{\,2}} = \frac{-\varepsilon}{\pi D} = - \frac{\Delta\lambda}{\lambda_2^{\,2}}$$

For the resonance at λ we have from $\frac{\Delta}{2} = \frac{2\pi D}{\lambda} = m\pi$

$D = \frac{m}{2}\lambda$. We introduce it and get

$$\frac{\varepsilon}{\pi \frac{m}{2} \lambda} = + \frac{\Delta \lambda}{{\lambda_2}^2} .$$

We write $\lambda_2 \simeq \lambda$ and take the absolute value (not indicated).

$$\frac{\lambda}{\Delta\lambda} = \frac{\pi m}{2\varepsilon} \qquad (2)$$

From (1) we have

$$\frac{1}{1 + g^2 m^2 \frac{\Delta}{2}} = \frac{1}{2} \qquad g^2 m^2 \frac{\Delta}{2} = 1$$

with $\frac{\Delta}{2} = \frac{2\pi}{\lambda} D = \frac{m\pi}{2} + \varepsilon$ we get

$$g^2 \sin^2(\frac{m\pi}{2} + \varepsilon) = g^2 \sin^2\varepsilon = 1$$

Since ε is small we have $g \cdot \varepsilon = 1$ (3)

Combining (2) and (3) we get

$$\frac{\lambda}{\Delta\lambda} = \frac{\pi m}{2} g = \frac{m}{2} \frac{2r}{1 - r^2} = \frac{\pi m r}{1 - r^2}$$

11. *Interference of White Light on a Thin Film (Soap Bubble).* Consider the formula of the transmitted light for the plane parallel plate with constant g,

$$I_t = \frac{1}{1 + g^2 \sin^2 \Delta/2}$$

The minima and maxima of I_t are obtained from various values of

$$\frac{\Delta}{2} = \frac{2\pi}{\lambda} nD \cos\theta_2$$

where θ_2 is the angle of the refracted light in the plate. If we take $n = 1.33$ (i.e., the refractive index of water in the visible spectrum), we have

$$\frac{\Delta}{2} = \pi 5.32 \frac{D}{\lambda} \cos\theta_2$$

Consider a wedge-shaped water film between two microscope slides of index $n = 1.5$. If light of various wavelengths is incident on a wedge-shaped film, we will see different colors for different thicknesses of the wedge, depending on the interference condition for maximum. Assume that $\theta_2 = 0$, and find the minimum thickness ($\neq 0$) of the portion of the wedge for the following:

a. $\lambda = 0.4\ \mu m$ (blue)
b. $\lambda = 0.5\ \mu m$ (green)
c. $\lambda = 0.7\ \mu m$ (red)

Solution

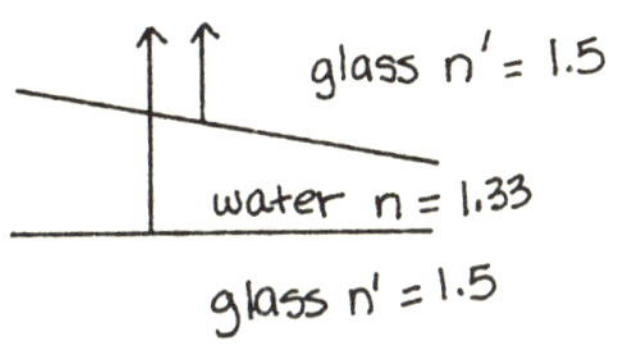

When $I_t = 1$, $\sin^2 \frac{\Delta}{2} = 0 \rightarrow \frac{\Delta}{2} = m\pi \qquad m = 0,1,2....$

There is a phase shift of π for the beam reflected from the second surface.

$$\frac{\Delta}{2} = \frac{2\pi}{\lambda}\, 2nD + \pi = m\pi$$

$$D = \frac{(m+1)\lambda}{4n}$$

For minimum thickness $m = 0$

(a) $\frac{(m+1)(0.4)}{4(1.33)} = 0.075\mu m$

(b) $\frac{(m+1)(0.5)}{4(1.33)} = 0.094\mu m$

(c) $\frac{(m+1)(0.7)}{4(1.33)} = 0.132\mu m$

12. *A Double Star and the Michelson Stellar Interferometer.* We want to use the stellar interferometer to estimate the angle ϕ between the two components of a double star. Assume that $\lambda = 0.5\ \mu m$ and $h = 100$ in. Calculate ϕ.

Solution

$$|\Delta h| = \frac{\lambda}{2\phi} \qquad \phi = \frac{0.5\mu}{2\cdot 100\text{in}}$$

$$\phi = \frac{0.5\cdot 10^{-4}\text{cm}}{2\cdot 250\text{cm}} = 0.1\cdot 10^{-6}\text{rad}$$

$$1\text{ rad} = 57.3° \qquad \phi = 5.73\cdot 10^{-6}\text{ degrees}$$

$$= 0.02\text{ sec.}$$

---

3 Diffraction

1. *Slit Aperture.* For the small angle approximation, the intensity was shown to be given by

$$u^2 = u_0^2\left(\frac{\sin[(2\pi/\lambda)(d/2)(Y/X)]}{(2\pi/\lambda)(d/2)(Y/X)}\right)^2$$

If we set $u_0^2 = 1$ at $Y = 0$, then the central maximum has unit intensity. Use the following numerical values: $d = 1$ mm, $\lambda = 500$ nm (visible region), and $X = 4$ m. Plot u^2 versus Y. Choose a scale such that $|u|^2 = 0.01$ corresponds to 1 cm and $Y = 1$ mm corresponds to 2 cm. Then, for graph paper with a grid of 10 lines per centimeter, the central maxima will be off scale, that is, off the upper end of the graph paper. (We are interested in the location of the side maxima.)

Solution

Slit aperture.

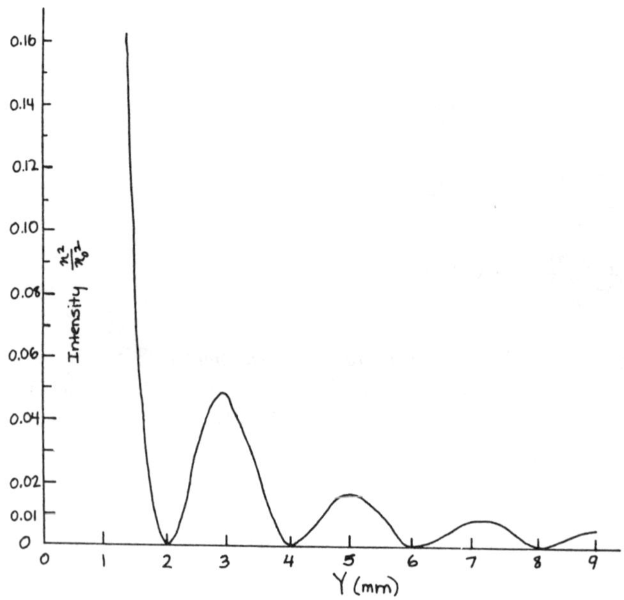

2. *Secondary Maxima of the Diffraction Pattern of a Slit.* We found the relation $b = \tan b$ [with $b = \pi(d/\lambda)(Y/X)$] for the location of the secondary maxima of a slit. Now using $\lambda = 500$ nm, $d = 1$ mm, $X = 4$ m, and Y in millimeters, graphically determine the values of the secondary maxima. Compare with plot of problem 1.

Solution

Secondary maxima of the diffraction pattern of a slit.

$$b = \tan b \text{ and } b = \left[\frac{\pi d}{\lambda} \frac{Y}{X} \right] = \frac{\pi Y}{2}$$

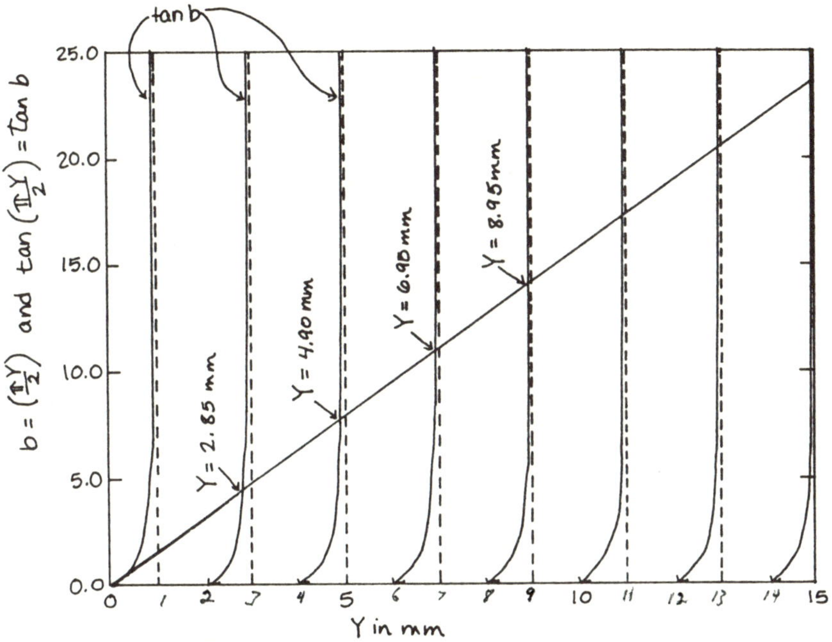

As Y increases the peaks come closer and closer to

$$b = (m + \tfrac{1}{2})\pi = \frac{\pi d}{\lambda} \frac{Y}{X} \Rightarrow Y = (m + \tfrac{1}{2}) \frac{X\lambda}{d}$$

in mm $\Rightarrow Y = (m + \frac{1}{2})2 = 3, 5, 7, \ldots$

3. *Double Slit: $d \ll a$.* Consider the double-slit experiment and assume that $d \ll a$.

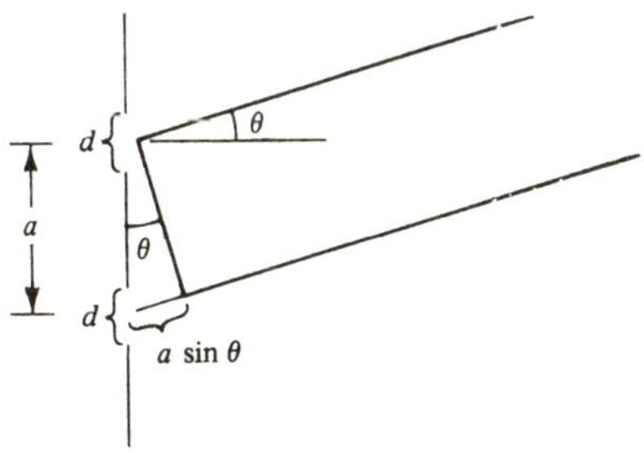

For the intensity of the diffraction pattern, we obtained

$$u^2 = 4\frac{\sin^2((2\pi/\lambda)\sin\theta(d/2))}{((2\pi/\lambda)\sin\theta(d/2))^2}\cdot\cos^2((2\pi/\lambda)\sin\theta(a/2))$$

The first factor describes the diffraction on the single slit; the second factor describes the interference effect. Derive the conditions for maxima (approximate) and minima that are equal to what we derived by elementary derivation.

<u>Solution</u>

Double slit

The special cases (using amplitudes) for peaks:

Single slit term for $\theta = 0$:

$$\left[\frac{\sin(\frac{2\pi}{\lambda}\{\sin\theta\}\frac{d}{2})}{\frac{2\pi}{\lambda}\{\sin\theta\}\frac{d}{2}}\right] \xrightarrow[\theta \to 0]{} \left[\frac{\frac{2\pi}{\lambda}\frac{d}{2}\cos\theta\{\cos[\frac{2\pi}{\lambda}\{\sin\theta\}\frac{d}{2}]\}}{\frac{2\pi}{\lambda}\{\cos\theta\}\frac{d}{2}}\right]_{\theta \to 0} \to 1 \quad \text{peak}$$

For the peaks of the single slit term the <u>peaks</u> are <u>approximately</u> at

$$\frac{2\pi d}{\lambda 2}\sin\theta = (2m+1)\frac{\pi}{2} \quad \text{where:} \quad m = 0,1,2,\ \ldots.$$

$(2m+1)\lambda = 2d\sin\theta$ <u>peak</u>

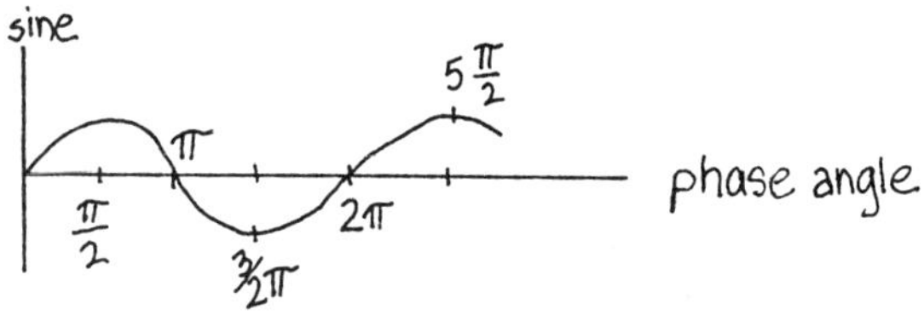

For the peaks of the interference term the peaks are at:

$$(\frac{2\pi}{\lambda}\{\sin\theta\}\frac{a}{2} = m\pi \quad \text{where:} \quad m = 0,1,2,\ \ldots.$$

$\Rightarrow m\lambda = a\sin\theta$ <u>peak</u>

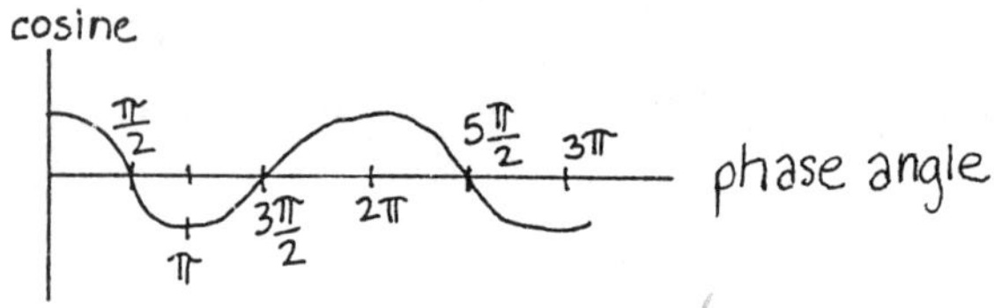

For the minima of the single slit term the peaks are at:

$$\frac{d}{2}\frac{2\pi}{\lambda}\sin\theta = m\pi$$

where: m = 1,2,3,...; the positive integers

rearranging:

$$m\lambda = d \sin\theta \qquad \underline{\underline{\text{minima}}}$$

The minima in the interference term:

$$\frac{a}{2}\frac{2\pi}{\lambda} \sin\theta = (2m+1)\frac{\lambda}{2}$$

where: m = 0,1,2,3,....

$$\Rightarrow (2m+1)\lambda = 2a \sin\theta \quad \underline{\underline{\text{minima}}}$$

4. *Double Slit: $a = 2d$.* Consider the double-slit experiment and assume that $a = 2d$

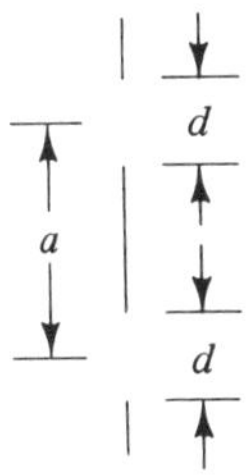

The intensity is given as

$$u^2 = 4\frac{\sin^2((2\pi/\lambda)\sin\theta(d/2))}{((2\pi/\lambda)\sin\theta(d/2))^2}\cdot\cos^2((2\pi/\lambda)\sin\theta(a/2))$$

a. Make a sketch of the diffraction factor, labeling the minima.
b. Make a sketch of the interference factor, labeling the minima and maxima.
c. Make a sketch of u^2 and indicate the "missing order."

Solution

(a)

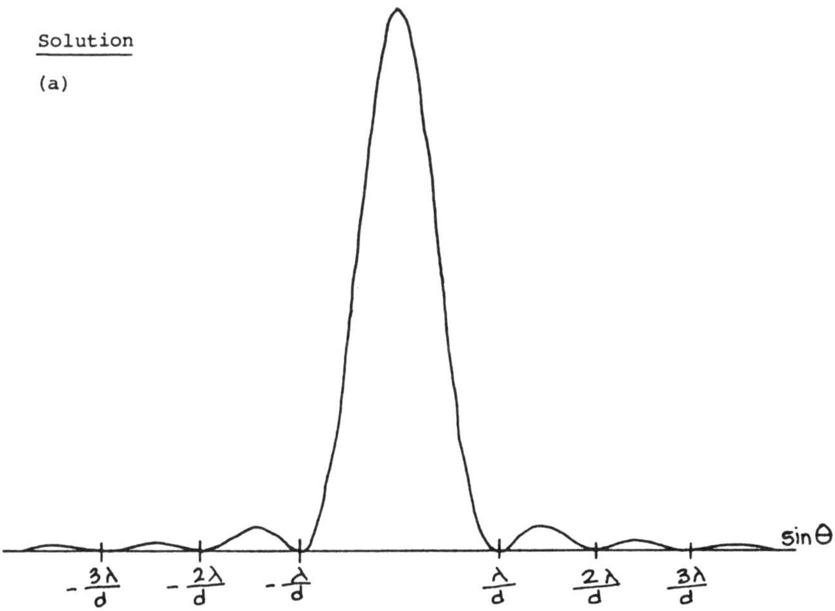

minima: $m\lambda = d \sin\theta$ m = 1,2,3

(b)

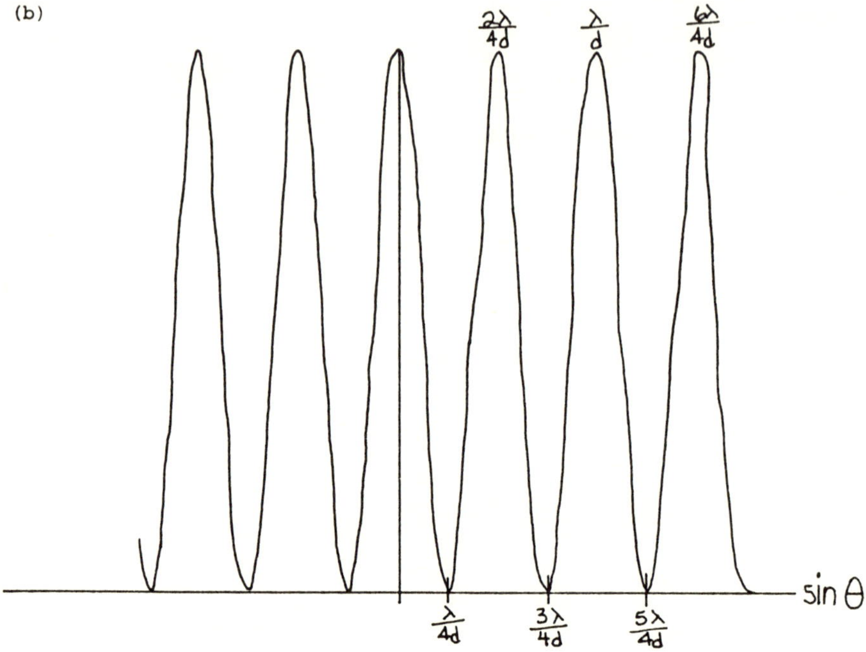

minima: $(2m+1)\lambda = 2a \sin\theta$ $m = 0,1,2,3$

$a = 2d$, $(2m+1)\lambda = 4d \sin\theta$

(c)

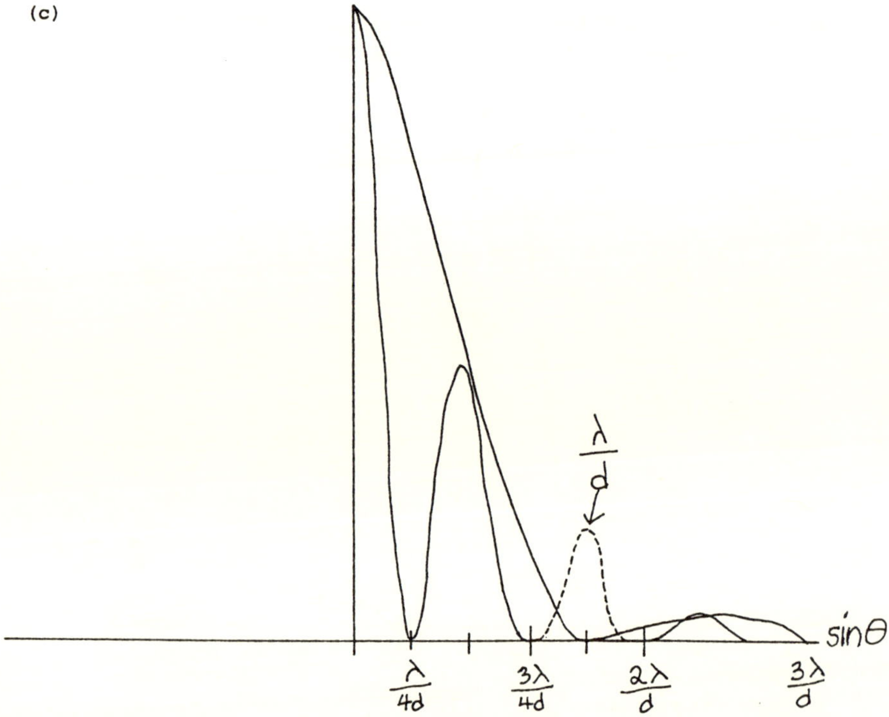

5. *Diffraction on a Double Slit for Two Different Wavelengths.* Consider the pattern of a double slit for two different wavelengths (not in the apparatus at the same time). For the intensity, we have

$$u^2 = u_0^2 \frac{\sin^2 \gamma(d/2)}{[\gamma(d/2)]^2} \cos 2\, \gamma(a/2)$$

Set $u_0^2 = 1$, $a = 5d$, $d = 2.5\ \mu$m, and $\gamma = (2\pi/\lambda)\sin\theta$. Plot the pattern as a function of θ in degrees.

a. For blue light, use $\lambda = 0.4\ \mu$m and find the maxima and minima for the two factors.

$\frac{\sin^2 \gamma(d/2)}{(\gamma(d/2))^2}$ 0 Max:
1 Min:
1 Max:
2 Min:
2 Max:

$\cos^2 \gamma\frac{a}{2}$ 0 Max:
1 Min:
1 Max:
2 Min:
2 Max:

b. Repeat the calculation for red light, where $\lambda = 0.7\ \mu m$.

c. Draw the two patterns on the same figure, one with a blue pencil, one with a red pencil.

<u>Solution</u>

Diffraction on a double slit for two different wavelengths

(not in the apparatus at the same time).

$$u_i^2 = \frac{\sin^2\{\frac{2.5}{2}\frac{2\pi}{\lambda_i}\sin\theta\}}{\{\frac{2.5}{2}\frac{2\pi}{\lambda_i}\sin\theta\}^2} \cos^2\{\frac{5d}{2}\frac{2\pi}{\lambda_i}\sin\theta\}$$

$$\Rightarrow \quad u_i^2 = \frac{\sin^2\{\frac{2.5\pi}{\lambda_i}\sin\theta\}}{\{\frac{2.5\pi}{\lambda_i}\sin\theta\}^2} \cos^2[\frac{12.5\pi}{\lambda_i}\sin\theta]$$

with λ_i in μm. i.e. $\gamma d/2 = \{\frac{2.5\pi}{\lambda_i}\sin\theta\}$ from $\gamma = \frac{2\pi}{\lambda}\sin\theta$

(a) <u>blue light</u> with $\lambda_1 = 0.40$ μm

From the text or problem 3 for the <u>single slit</u> term: $(\sin^2 q/q^2)$

$$\begin{cases} \text{peaks:} & \theta=0 \text{ plus } \sin\theta = \frac{(2m+1)\lambda}{2d} = (2m+1)0.08 \\ & m = 0,1,2,\ldots \\ \text{minima:} & \sin\theta = m\lambda/d = \frac{m(0.4)}{2.5} = 0.16m \\ & m = 1,2,3,\ldots \end{cases}$$

m	extreme	θ (deg)	$\frac{\sin^2(\gamma\frac{d}{2})}{(\gamma d/2)^2}$
0	max.	4.59	0.406
1	min.	9.20	0
1	max.	13.9	0.0451
2	min.	18.7	0
2	max.	23.6	0.0162

plus a Max. at $\theta = 0°$ equal to unity

$$\gamma \frac{d}{2} = \frac{2.5\pi}{0.4} \sin\theta = (19.6 \sin\theta) \text{ rad} = (1123 \sin\theta)°$$

<u>blue light</u> for the <u>multiple slit</u> term ($\cos^2 p$)

peaks: $\sin\theta = \frac{m\lambda}{a} = \frac{m0.4}{12.5} = 0.032m$ where $m = 0,1,2,....$

minima: $\sin\theta = \frac{(2m+1)\lambda}{2a} = \frac{(2m+1)0.4}{25} = 0.016(2m+1)$

note: $\gamma = \frac{2\pi}{\lambda} \sin\theta$ where: $m = 0,1,2,....$

So: $\frac{\gamma a}{2} = \frac{12.5\pi}{0.4} \sin\theta = (98.2 \sin\theta)\text{rad} = (5625 \sin\theta)°$

m	extreme	sin θ	θ°	$\cos^2[\gamma\frac{a}{2}]$
0	Min.	0.016	0.917	0
0	Max.	0	0	1
1	Min.	0.048	2.75	0
1	Max.	0.032	1.83	1
2	Min.	0.080	4.59	0
2	Max.	0.064	3.67	1

(b) <u>red light</u> with $\lambda_2 = 0.7\mu m$

<u>single slit</u> term $(\sin^2 q)/q^2$

peaks: $\theta = 0$ plus $\sin\theta = \frac{(2m+1)\lambda}{2d} = (2m+1)0.14$ where $m = 0,1,2,...$

minima: $\sin\theta = \frac{m\lambda}{d} = 0.28m$ where $m = 1,2,3,....$

$$\gamma d/2 = \frac{2.5\pi \sin\theta}{0.7} = (11.22 \sin\theta)\text{rad.} = (643 \sin\theta) \text{ degrees}$$

m	extremes	$\sin\theta$	$\theta°$	$\frac{\sin^2(\gamma d/2)}{(\gamma d/2)^2}$
0	Max.	0.14	8.05	0.407
1	Min.	0.28	16.3	0
1	Max.	0.42	24.8	0.0450
2	Min.	0.56	34.1	0
2	Max.	0.70	44.4	0.0162

plus a max, at $\theta = 0$ with amplitude of unity

red light with $\lambda = 0.7\mu m$; multiple slit term ($\cos^2 p$)

peaks: $\sin\theta = \frac{m\lambda}{a} = 0.056\ m$ where: $m = 0,1,2,\ldots.$

minima: $\sin\theta = \frac{(2m+1)\lambda}{2a} = 0.028(2m+1)$ where $m = 0,1,2,\ldots.$

Note $\gamma = \frac{2\pi}{\lambda}\sin\theta \Rightarrow (\frac{\gamma a}{2}) = \frac{12.5\pi}{0.7}\sin\theta = (56.1\sin\theta)\text{rad}$.
$= (3214\sin\theta)°$

m	extreme	$\sin\theta$	$\theta°$	$\cos^2(\gamma\frac{a}{2})$
0	Min.	0.028	1.60	0
0	Max.	0	0	1
1	Min.	0.084	4.82	0
1	Max.	0.056	3.21	1
2	Min.	0.140	8.05	0
2	Max.	0.112	6.43	1

(c)

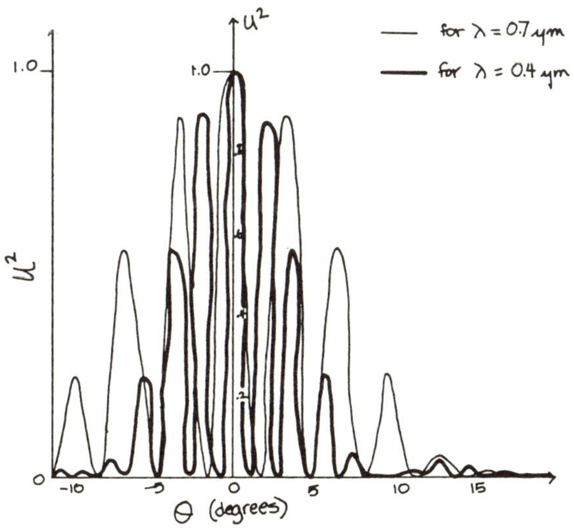

6. *Diffraction on a Plane Grating: Oblique Incident Light.* Consider the diffraction on a plane grating with periodicity constant a and slit opening d. Assume that the incident light makes the angle ϕ with respect to the normal of the grating. The diffraction angle θ is as shown in the figure.

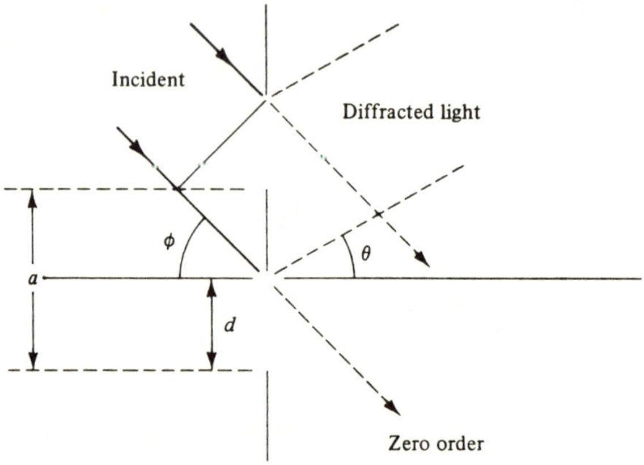

a. Derive the intensity formula for the diffracted light. The result must be the same as in equation 3.36 for $\phi = 0$.
b. Show that for the choice of $\sin\theta + \sin\phi = m(\lambda/a)$, the interference factor is equal to N^2.
c. Calculate the ratio of the intensities for $m = 1$ with respect to $m = 0$ for the case where $a = 8d$.

Solution

Diffraction on a plane grating. Oblique incident light.

(a)

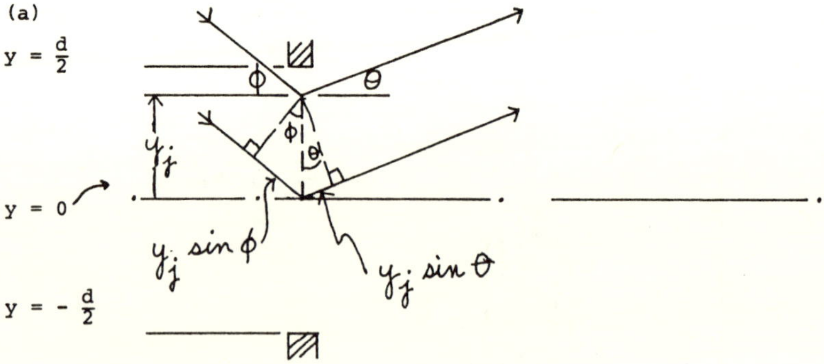

The phase distances are no longer just $y_j \sin\theta$ but now are $y_j (\sin\theta + \sin\phi)$.

Recall for $\phi = 0$, from Eq. (3.9) we note ($C^2 = 1$)

$$u^*u = \frac{\sin^2 (k \frac{d}{2} \sin\theta)}{(k\frac{d}{2} \sin\theta)^2}$$

The limits of integration were $-\frac{d}{2}$ to $+\frac{d}{2}$ while the variable was y. Thus with the same limits we expect the factor d/2 unchanged. However, replace sinθ with (sinθ + sinϕ). Thus we have

$$\text{one aperture:} \qquad u^*u\Big|_{(1\ \text{slit})} = \frac{\sin^2[k\frac{d}{2}(\sin\theta + \sin\phi)]}{[k\frac{d}{2}(\sin\theta + \sin\phi)]^2}$$

$$\text{N apertures:} \qquad u^*u\Big|_{(N\ \text{slits})} = \left\{\frac{\sin^2[Nk\frac{a}{2}(\sin\theta + \sin\phi)]}{\sin^2[k\frac{a}{2}(\sin\theta + \sin\phi)]}\right\}$$

$$u^*u\Big|_{\text{overall}} = \left\{\frac{\sin^2[k\frac{d}{2}(\sin\theta + \sin\phi)]}{[k\frac{d}{2}(\sin\theta + \sin\phi)]^2} \quad \frac{\sin^2[Nk\frac{a}{2}(\sin\theta + \sin\phi)]}{\sin^2[k\frac{a}{2}(\sin\theta + \sin\phi)]}\right\}$$

(b) Set $(\sin\theta + \sin\phi) = \frac{m\lambda}{a}$

This comes from:

$$\frac{ka}{2}(\sin\theta + \sin\phi) = m\pi$$

where $m = 1,2,3,\ldots$

We have to use l'Hopital's rule on the multiple slit term amplitudes: [taking derivatives w.r.t (sinθ + sinϕ)]

$$\frac{[Nk\frac{a}{2}\cos\{Nk\frac{a}{2}(\sin\theta + \sin\phi)\}}{k\frac{a}{2}\cos\{k\frac{a}{2}(\sin\theta + \sin\phi)\}}$$

$$= N\frac{\cos\{Nm\pi\}}{\cos\{\pi\}} = N$$

$$= \quad u^*u = N^2\frac{\sin^2[\frac{kd}{2}(\sin\theta + \sin\phi)]}{[\frac{kd}{2}(\sin\theta + \sin\phi)]^2}$$

$$= N^2\frac{\sin^2\{m\pi\frac{d}{a}\}}{\{m\pi\frac{d}{a}\}^2}$$

(c) Set $a = 8d$ and $m = 1$

$$u^*u = N^2\frac{\sin^2(\pi/8)}{(\pi/8)^2} = N^2\frac{0.146}{0.154} = 0.948\ N^2$$

$$\text{ratio} \quad \frac{u^*u(m=1)}{u^*u(m=0)} = .948$$

Since the peak is very near the maximum, i.e. N^2, a great deal of energy is in the m = 1 diffraction peak and direction.

7. *Circular Aperture.* An important practical example is the diffraction pattern due to a circular aperture. For example, a circular lens interrupting a wave produces diffraction, and the resolving power of the lens is limited by diffraction considerations. This application provides us with a chance to use Bessel functions. Using the following coordinate system and, in analogy with the rectangular aperture, consider the integral

$$u = u_0 \int_{\substack{\text{over}\\ \text{opening}}} \exp\left\{ i2\pi\left[\left(\frac{y}{\lambda}\right)\left(\frac{Y}{X}\right) + \left(\frac{z}{\lambda}\right)\left(\frac{Z}{X}\right)\right]\right\} dy\, dz$$

Since rectangular coordinates are not appropriate for circular geometry, the polar coordinates shown in the figure will be used instead.

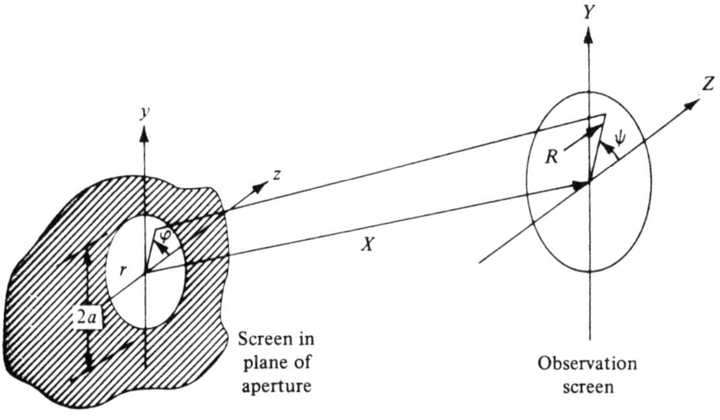

These variables satisfy the following transformation equations:

$$y = r\sin\phi \qquad Y = R\sin\psi$$
$$z = r\cos\phi \qquad Z = R\cos\psi$$

where $0 \le r \le a$, $0 \le \phi \le 2\pi$, and $0 \le \psi \le 2\pi$.

a. Write $[(y/\lambda)(Y/X) + (z/\lambda)(Z/X)]$ and the area element $dy\,dz$ in coordinates r, R, ϕ, and ψ and obtain the following integral:

$$u = u_0 \int_{-\pi}^{+\pi}\int_0^a e^{i2\pi(r/\lambda)(R/X)\cos(\psi-\phi)} r\, dr\, d\phi$$

b. Rewrite the integral by using the definition of the Bessel function of order zero

$$J_0(q) = \frac{1}{2\pi}\int_{-\pi}^{+\pi} e^{iq\cos(\phi-\psi)}\, d\phi$$

and arrive at

$$u = u_0\left(\frac{\lambda X}{2\pi R}\right)^2 2\pi \int_{q=0}^{q=2\pi(aR/\lambda X)} J_0(q) q\, dq$$

Provided is a graph of J_0 and J_1 as a function of $q(1 \to 18)$.

c. To carry out the integration, use an identity for Bessel functions relating the zeroth order $J_0(q)$ and the first order $J_1(q)$:

$$\int_0^{q'} J_0(q) q\, dq = q' J_1(q')$$

(′ is not a derivative). Arrive at the following:

$$u(R) = u_0 2\pi a^2 \frac{J_1[2\pi(a/\lambda)(R/X)]}{2\pi(a/\lambda)(R/X)}$$

d. Find the value of $u^2(R)$ for $R = 0$. To do this, consider the series expansion of $J_1(q)$

$$J_1(q) = \frac{q}{2}\left[1 - \frac{1}{2}\left(\frac{q}{2}\right)^2 + \frac{1}{12}\left(\frac{q}{2}\right)^4 + \cdots\right]$$

If you sct $R = 0$, you will get 0/0. Application of L'Hôpital's rule will give you the desired result:

$$\lim_{R\to 0} |u|^2 = 1$$

e. Find the zeros of $J_1(q)$ from the graph (or tables you have access to), and give the position of the zero of the function in R/X.

	R/X
1 zero $q =$	
2 zero $q =$	
3 zero $q =$	

The first minimum is located at $R/X = 0.61\lambda/a$ or, if we introduce the diameter $\boldsymbol{d}$ of the aperture, at $R/X = 1.22\lambda/\boldsymbol{d}$. This relationship is used in the discussion of the resolution of optical instruments. The secondary maxima, as was found in the single-slit case, are almost, but not exactly, located between the minima points. The central maximum, containing more than 80 percent of the total diffracted energy, is called the Airy disk and corresponds to the region $|R/X| < 3.83\lambda/2\pi a$. This central maximum is surrounded by concentric rings, which reduce in intensity as R/X increases.

Intuitively, you might expect the circular aperture diffraction pattern to resemble that of the slit pattern when the latter is rotated about its central maximum. Indeed, this is observed. Quantitatively, however, the intensities of the secondary maxima of the circular aperture pattern are less than their single-slit counterparts. Also, the positions of the minima are further from the central maxima for the circular aperture than for the slit.

f. Make a plot of the diffraction patterns of a slit, using (see problem 1.)

$$u^2 = u_0^2\left(\frac{\sin(2\pi/\lambda)(d/2)(Y/X)}{2\pi/\lambda(d/2)(Y/X)}\right)^2$$

For $u_2^0 = 1$, $d = 1$ mm, $\lambda = 500$ nm, and $X = 4$ m. Plot u^2 versus Y, and choose a scale such that $|u|^2 = 0.01$ corresponds to 1 cm on the graph paper and $Y =$ 1 mm corresponds to 2 cm on the paper. (For typical graph paper, the central maximum will be off scale, but we are not interested in that position.)

g. On the same graph paper, plot the circular aperture, using

$$u^2 = 4\left(\frac{J_1(2\pi)(a/\lambda)(R/X)}{(2\pi)(a/\lambda)(R/X)}\right)^2$$

Take $2a = d = 1$ mm, $\lambda = 500$ nm, and $X = 4$ m. Read off your graph the first two minima for both cases and the first secondary maxima (between them) and intensities, and fill in the Table.

	First minima	Intensity	Next maxima	Intensity	Second minima	Intensity
Slit						
Round aperture						

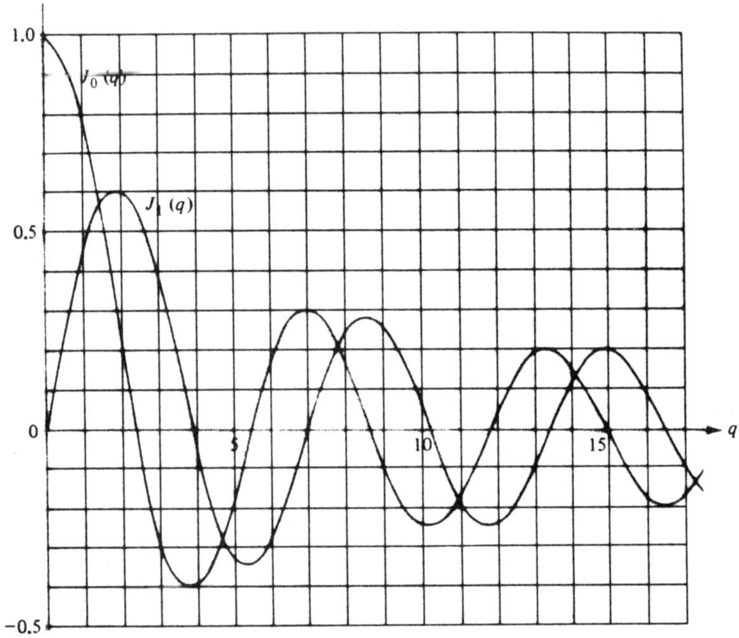

$J_0(q)$ and $J_1(q)$

Solution

(a) $\left(\frac{y}{\lambda}\frac{Y}{X} + \frac{z}{\lambda}\frac{Z}{X}\right) = \left(\frac{r\sin\phi}{\lambda}\frac{R\sin\psi}{X} + \frac{r\cos\phi}{\lambda}\frac{R\cos\psi}{X}\right)$

$= \frac{rR}{\lambda X}\ (\sin\phi\ \sin\psi + \cos\phi\ \cos\psi)$

From trigonometry: $\cos(A-B) = \cos A \cos B + \sin A \sin B$

substituting:

$\left(\frac{y}{\lambda}\frac{Y}{X} + \frac{z}{\lambda}\frac{Z}{X}\right) = \left(\frac{r}{\lambda}\frac{R}{X}\cos\,[\psi-\phi]\right)$

$dA \equiv dydz$ = area in rectilinear coordinates; in circular coordinates:

$dA = dSdr = rd\phi dr$

Combining

$$u = u_o \int_{-\pi}^{+\pi} \int_o^a e^{i2\pi \frac{r}{\lambda} \frac{R}{X} \cos(\psi-\phi)} rdrd\phi$$

Since we're dealing with an even function i.e. $\cos(\psi-\phi)$ the limits of integration can be $-\pi$ to $+\pi$ as well as from 0 to 2π for ϕ.

(b) Given the zeroth order Bessel function:

$$J_o(q) \equiv \frac{1}{2\pi} \int_{-\pi}^{+\pi} e^{iq \cos(\phi-\psi)} d\phi$$

comparing with part (a) let:

$$q \equiv 2\pi\frac{r}{\lambda} \frac{R}{X} \quad \Rightarrow \quad dq = \frac{2\pi}{\lambda} \frac{R}{X} dr$$

$$\Rightarrow \quad qdq = \frac{2\pi}{\lambda} \frac{R}{X} \frac{2\pi}{\lambda} \frac{R}{X} rdr = \frac{4\pi^2}{\lambda^2} \frac{R^2}{X^2} rdr$$

$$\Rightarrow \quad rdr = \left(\frac{\lambda X}{2\pi R}\right)^2 qdq$$

Combining with part (a):

$$u = u_o (2\pi) \left(\frac{\lambda X}{2\pi R}\right)^2 \int_o^{2\pi aR/(\lambda X)} J_o(q)qdq$$

(c) We are given for the zeroth and first order Bessel function:

$$\int_o^{q'} J_o(q)qdq = q' J_1(q')$$

Comparing with the integral in part (b):

$$u = u_o(2\pi)\left(\frac{\lambda X}{2\pi R}\right)^2 \left(\frac{2\pi aR}{\lambda X}\right) J_1\left(\frac{2\pi aR}{\lambda X}\right)$$

$$u = u_o (2\pi a^2) \left\{ \frac{J_1\left(\frac{2\pi aR}{\lambda X}\right)}{\left(\frac{2\pi aR}{\lambda X}\right)} \right\}$$

(d) $J_1(q) = \frac{q}{2} [1 - \frac{1}{2} (\frac{q}{2})^2 + \frac{1}{12} (\frac{q}{2})^4 + ...]$

Note: $[\frac{J_1(q)}{q}] \rightarrow \frac{0}{0}$ for $R \rightarrow 0$

Using l'Hopital's rule (see in "Note" below)

$$\left\{ \frac{J_1(q)}{q} \right\} \underset{R\to 0}{\rightarrow} \frac{\frac{d}{dq}[J_1(q)]}{\frac{d}{dq}(q)} = \frac{\frac{1}{2} - \frac{1}{16} 3q^2 + ...}{1} \xrightarrow[R\to 0]{} \frac{1}{2}$$

at the origin i.e. R=0, one has

$$u \underset{R\to 0}{\to} u_o \cdot (2\pi a^2) \cdot \frac{1}{2} \quad \text{and} \quad u^2 \underset{R\to 0}{\to} \frac{1}{4}\,(2\pi a^2)^2 \cdot u_o^2$$

Note: If $\left(\frac{f(X)}{f(X)}\right)^2 \to \frac{0}{0}$ where f and g are arbitrary functions

$$\text{Then } \frac{\frac{d}{dX}\, f^2(X)}{\frac{d}{dX}\, g^2(X)} \to \frac{2\; f(X)\;\; f'(X)}{2\; g(X)\;\; g'(X)} \to \frac{0}{0}$$

{ where in this note ' is a derivative. }

The L'Hospital again:

$$\frac{[f'(X)]^2 + ff''}{[g'(X)]^2 + gg''} \to \left(\frac{\frac{df}{dX}}{\frac{dg}{dX}}\right)^2$$

i.e. only need the 1st derivatives squared:

If $u_o^2 = \dfrac{4}{(2\pi a^2)^2}$ then $\dfrac{u^2(R\to 0)}{u_o^2} = 1$

or

$$u^2(R) = \frac{4J_1^2\left(\frac{2\pi aR}{\lambda X}\right)}{\left(\frac{2\pi aR}{\lambda X}\right)^2} \xrightarrow[R\to 0]{} 1$$

$$\xi \equiv \frac{aR}{\lambda X} \qquad \frac{R}{X} = \frac{\lambda}{a}\,\frac{1}{2\pi}\,q$$

	zeros in $J_1(q)$	$\frac{R}{X}$
1.	q = 3.9	0.62
2.	q = 7.0	1.11
3.	q = 10.1	1.61

For the 1st minimum

=> $\frac{R}{X} = 0.61\frac{\lambda}{a} = 1.22\frac{\lambda}{D}$

where: D≡(diameter of hole)

f) and g)

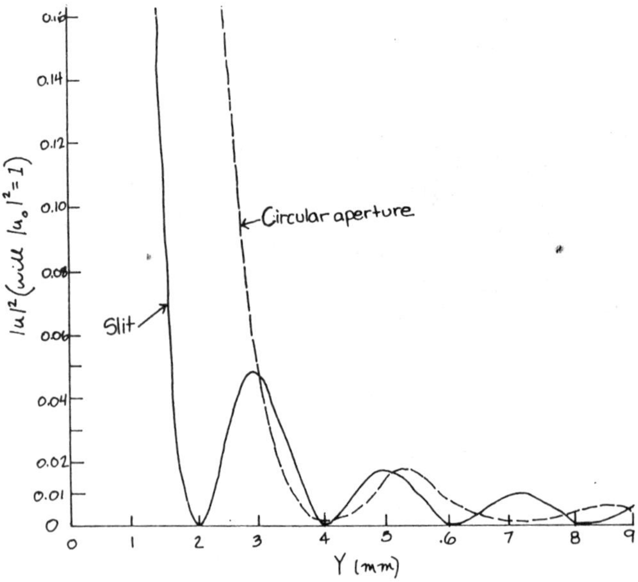

	First Min.	Intensity	Next Maxima	Intensity	Second Minima	Intensity
Slit	2.0	0	3.0	0.046	4.0	0
Round Aperture	2.48	0	3.39	0.02	4.46	0

8. *Wave Equation.* Show that $u(r) = e^{ikr}/r$ is a solution of the wave equation $\nabla^2 u + k^2 u = 0$.

<u>Solution:</u> In spherical polar coordinates we have

$$\nabla^2 = \frac{1}{r^2} \frac{\partial}{\partial r} (r^2 \frac{\partial}{\partial r}) + \text{terms in the angles } \theta \text{ and } \phi$$

Application of ∇^2 to u(r)

$$\nabla^2 u(r) = \frac{1}{r^2} \frac{\partial}{\partial r} [r^2 (\frac{\partial}{\partial r} \frac{e^{ikr}}{r})]$$

$$= \frac{1}{r^2} \frac{\partial}{\partial r} [r^2 (\frac{ik}{r} - \frac{1}{r^2}) e^{ikr}]$$

$$= \frac{1}{r^2} \frac{\partial}{\partial r} (ikre^{ikr} - e^{ikr})$$

$$= \frac{1}{r^2} (ike^{ikr} + r(ik)^2 e^{ikr} - ike^{ikr})$$

$$= \frac{1}{r} ((ik)^2 e^{ikr}) = -k^2 \frac{e^{ikr}}{r} = -k^2 u(r)$$

9. *Fresnel Diffraction, Slit, Table.* Use Table 3.1 and plot the Fresnel diffraction pattern of a slit. Use $d = 1$ mm, $\lambda = 500$ nm, $X = 4$ m, and Y in millimeters from $Y = 0$ to 5 mm in steps of 0.5 mm.

<u>Solution</u>

Y_{mm}	η_1	$C(\eta_1)$	$S(\eta_1)$	η_2	$C(\eta_2)$	$S(\eta_2)$	$[C(\eta_2)-C(\eta_1)]^2$	$[S(\eta_2)-S(\eta_1)]^2$	Tot
0	.5	.492	.065	-.5	-.492	-.065	.968	.017	.985
.5	1.0	.780	.438	0	0	0	.608	.192	.80
.0	1.5	.445	.698	.5	.492	.065	.002	.401	.403
.5	2.0	.488	.343	1.	.780	.438	.085	.009	.094
	2.5	.457	.619	1.5	.445	.698	.000	.006	.006
.5	3	.606	.496	2	.488	.343	.014	.023	.037
	3.5	.533	.415	2.5	.457	.619	.006	.042	.048
.5	4	.498	.420	3	.606	.496	.012	.006	.018
	4.5	.526	.434	3.5	.533	.415	.000	.000	0
.5	5	.564	.499	4	.498	.420	.004	.006	.01
	5.5	.478	.554	4.5	.526	.434	.002	.014	.016

Fresnel diffraction from a slit

$$d = 1\text{mm} = 10^{-3}\text{m}, \qquad \lambda = 500\text{nm} = 5\cdot 10^{-7}\text{m}$$

$$X = 4\text{m}, \qquad \sqrt{\frac{2}{\lambda X}} = 10^3$$

$$\eta_1 = (Y_{mm} + 0.5)\, 10^{-3}\cdot 10^3 = Y_{mm} + 0.5$$

$$\eta_2 = (Y_{mm} - 0.5)$$

Division by .985 will result in the normalized pattern.

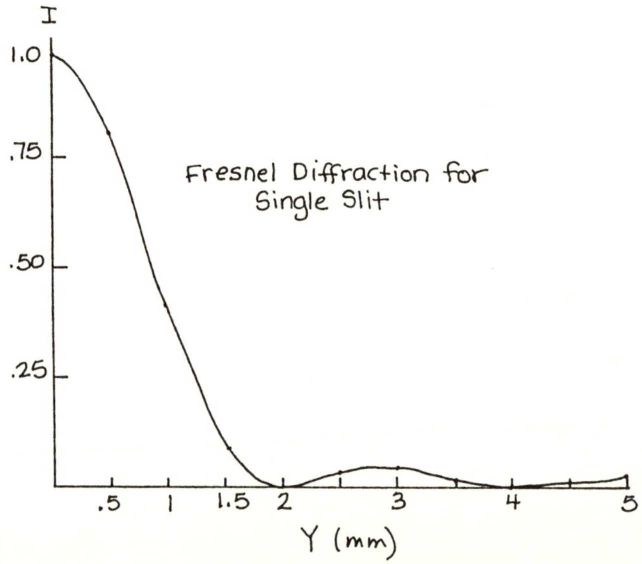

10. *Fresnel Diffraction, Slit, Cornu Spiral.* Plot the Fresnel diffraction pattern of a single slit by using the Cornu spiral as provided in the Solutions Manual. Use the following values: $d = 1$ mm, $\lambda = 500$ nm, $X = 4$ m, and Y from 0 to 4.5 mm in increments of 0.25 mm. Use

$$u = C\int_{\eta_1}^{\eta_2} e^{i(\pi/2)\eta^2}\, d\eta$$

and equations 3.71. The distance on the Cornu spiral between the parameters η_1 and η_2 is

$$\frac{u}{u_0} = \sqrt{[C(\eta_2) - C(\eta_1)]^2 + [S(\eta_2) - S(\eta_1)]^2} = \sqrt{\frac{I}{I_0}}$$

To normalize the intensity to unity at $Y = 0$, find the values of η_1 and η_2 corresponding to $Y = 0$. Fill in the table for η_1 and η_2. Note that the diffraction intensity is never 0.

For the plot, use graph paper in centimeter divisions and scale u^2 and Y as indicated below (see also Section 5.B):

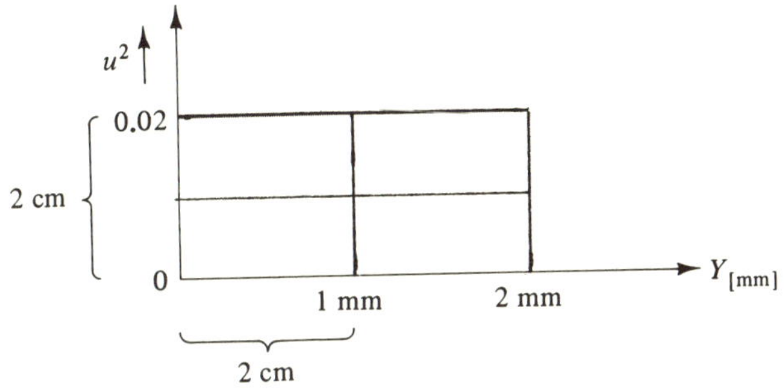

Solution (Cornu spiral is supplied on pages 68-69 of this manual.)

We have

$$\frac{u}{u_o} = \sqrt{[C(\eta_2) - C(\eta_1)]^2 + [S(\eta_2) - S(\eta_1)]^2} = \sqrt{\frac{I}{I_o}}$$

For $d = 1mm$ $\lambda = 500nm$ $X = 4m$

$$\eta_1 = (Y + \frac{d}{2})\sqrt{\frac{2}{\lambda X}} = (Y + 0.5)\sqrt{\frac{2}{(5 \times 10^{-7})(4)m^2}}$$

$$= (Y + 0.5mm)\ 10^3 \frac{1}{m} = (Y + 0.5), (Ymm)$$

$$\eta_2 = (Y - \frac{d}{2})\sqrt{\frac{2}{\lambda X}} = (Y - 0.5) \quad 10^3 \quad = (Y - 0.5), (Ymm)$$

Normalization $(Y = 0)$: $\eta_1 = 0.5$ $\eta_2 = -0.5$

Then

I_o is (distance between $\eta_1 = 0.5$ and $\eta_2 = -0.5$ on Cornu spiral)^2

$$= (18cm)^2 = (324cm)^2$$

Y_1mm	η_1	η_2	"distance"	("dist.")^2	Normalized Intensity
0	0.5	-0.5	18.0	324.0	1.000
0.5	1	0	16.6	276.3	0.853
1	1.5	0.5	11.7	136.0	0.420
1.5	2	1	5.7	32.4	0.100
2	2.5	1.5	1.5	2.2	0.007
2.25	2.75	1.75	2.5	6.1	0.019
2.5	3	2	3.4	11.5	0.036
2.75	3.25	2.25	4.2	17.8	0.055
3	3.5	2.5	3.9	15.6	0.048
3.25	3.75	2.75	3.5	12.2	0.038
3.5	4	3	2.4	5.7	0.018
3.75	4.25	3.25	1.6	2.4	0.008
4	4.5	3.5	0.6	0.3	0.001
4.25	4.75	3.75	1.7	2.9	0.009
4.5	5	4	1.8	3.4	0.010

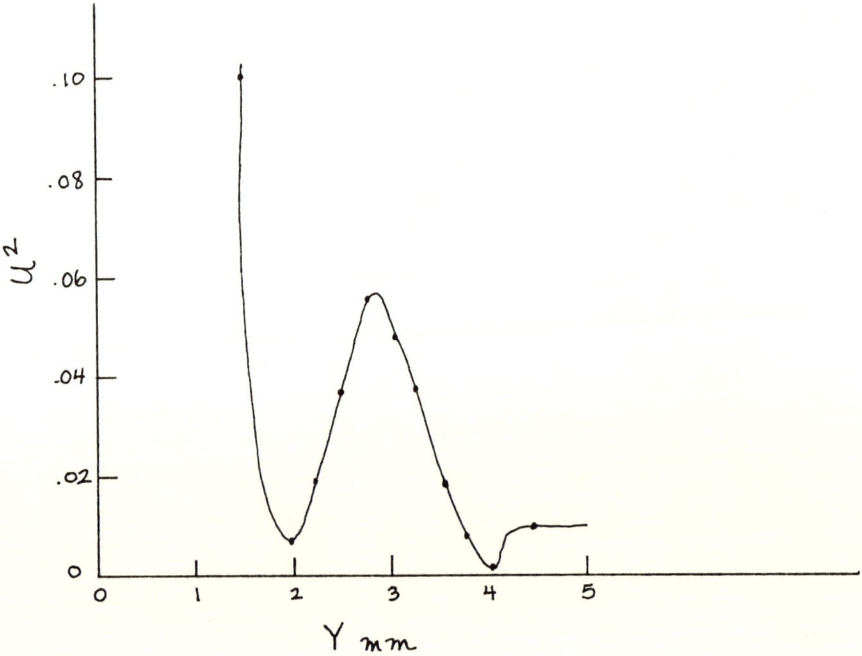

11. *Fresnel Diffraction, Half-Plane.* Plot the diffraction pattern obtained in the illuminated half-plane. Use the values $\lambda = 500$ nm and $X = 4$ m, and locate the half-plane such that the projection of the edge onto the screen is at $Y = 0$. Normalize the pattern to unity for $Y \to \infty$ (in the illuminated area; see Section 5B). Choose values of Y from $+2$ mm to -5 mm in increments of 1 mm. Compare Figure 3.40 to check if all minima and maxima are found.

Solution

To generate the edge pattern from the single slit formulation, take $-\frac{d}{2} = 0$ and $\frac{d}{2} = \infty$

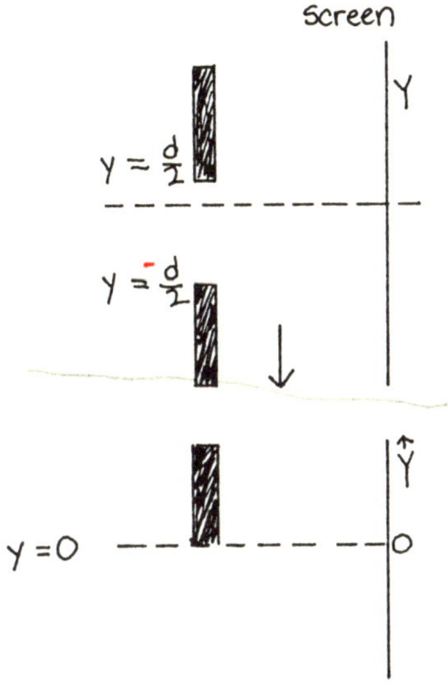

Then we have

$$\eta_1 = (Y+\tfrac{d}{2})\sqrt{\frac{2}{\lambda X}} \xrightarrow{-\frac{d}{2}=0} \eta_1 = Y\sqrt{\frac{2}{\lambda X}}$$

and

$$\eta_2 = (Y-\tfrac{d}{2})\sqrt{\frac{2}{\lambda X}} \xrightarrow{+\frac{d}{2}\to\infty} \eta_2 = -\infty$$

To normalize

$$u_1{}^2 = u_o{}^2 \{[C(0) - C(-\infty)]^2 + [S(0) - C(-\infty)]^2\}$$

$$u_1{}^2 = u_o{}^2 \ (9\text{cm})^2$$

$$u_1{}^2 = 81\ u_o{}^2 \qquad u_o{}^2 = \frac{1}{81}, \quad u_1{}^2 \text{ is } 1 \text{ for } \eta_1 = 0$$

Then

$$u_1{}^2(\eta_1) = \frac{[\text{distance from } \eta_1 \text{ to } -\infty \text{ on Cornu}]^2}{81}$$

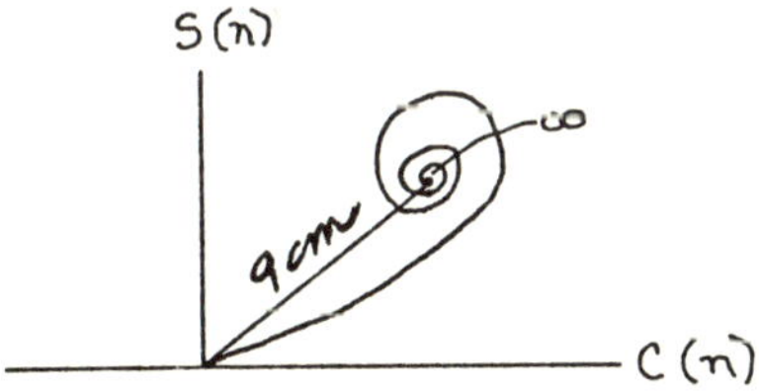

$\eta_1 = Y$	"distance"	("dist.")^2	Intensity
-3	1.3	1.6	0.02
-2	2.1	4.3	0.05
-1	3.6	12.9	0.16
0	9.0	81.0	1.0
1	20.0	398.0	4.9
1.25	20.8	431.4	5.3
1.9	15.1	228.2	2.8
2.25	19.4	378.2	4.7
2.75	16.4	267.8	3.3

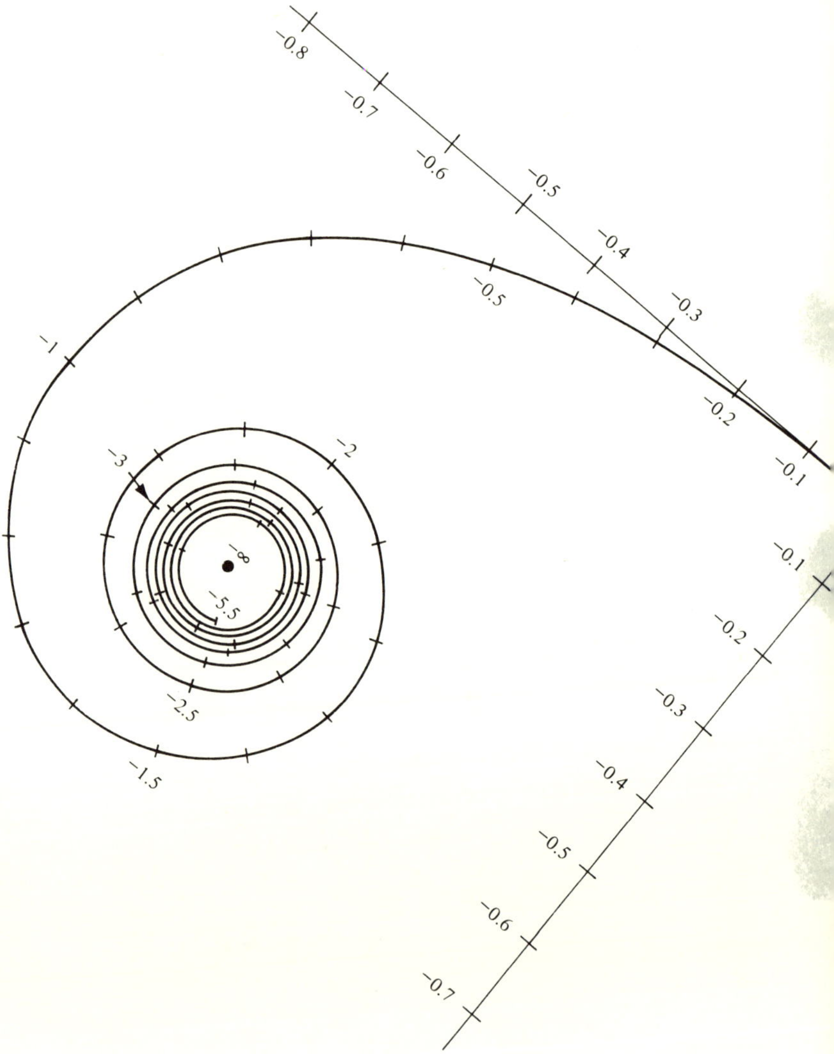

−0.8
−0.7
−0.6
−0.5
−0.4
−0.3
−0.2
−0.1
−0.5
−1
−2
−3
−8
−5.5
−2.5
−1.5
−0.1
−0.2
−0.3
−0.4
−0.5
−0.6
−0.7

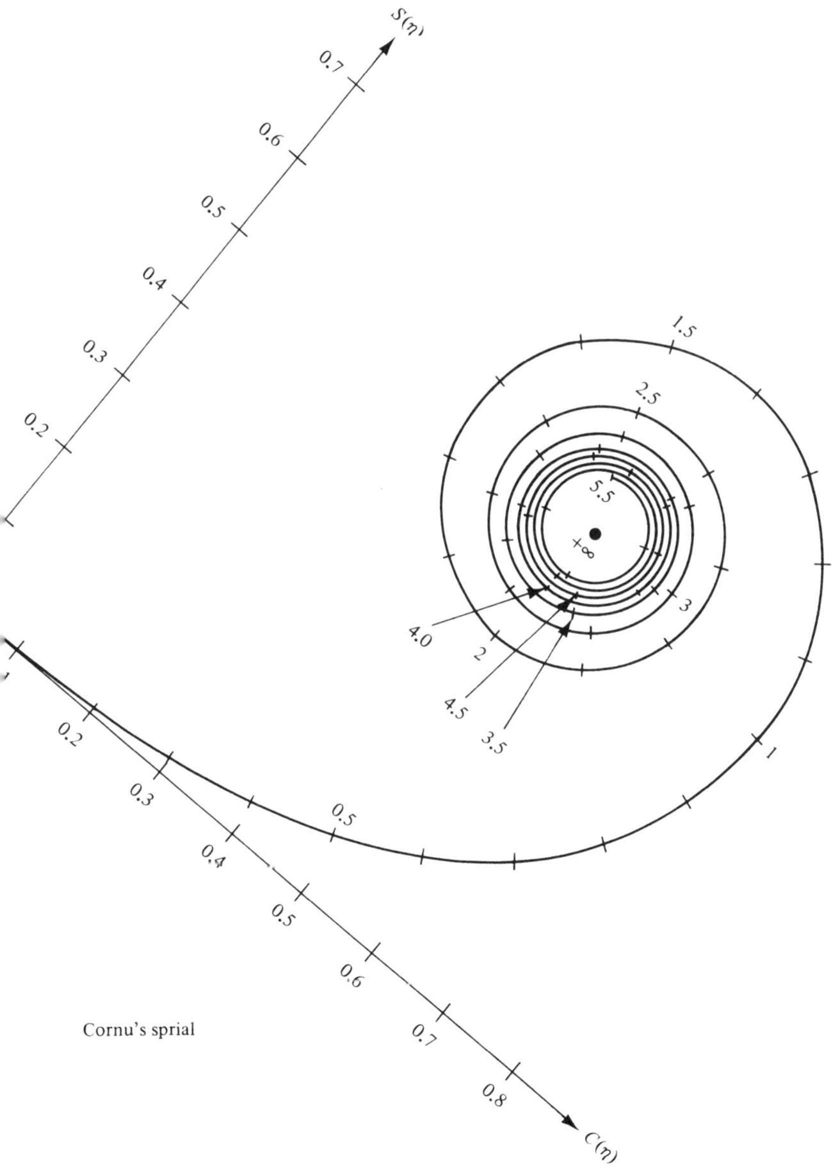

Cornu's sprial

Solution (Contd.)

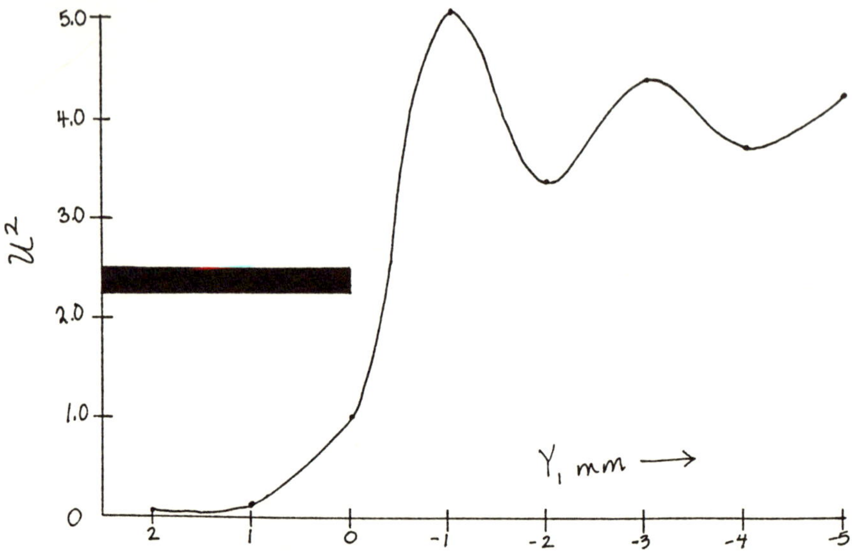

12. Sample calculation

Our starting point is an identity known as the divergence theorem which we write for any vector $\vec{F}$ as follows

$$\int_S \vec{F} \cdot \vec{n}\, d\sigma = \int_V \nabla \cdot \vec{F}\, dv \qquad (1)$$

The integral on the left is a surface integral (two-dimensional) and on the right, a volume (three-dimensional) integral. The quantity $\hat{n}$ represents a unit normal vector, that is a vector of unit length which is always normal to the surface S under consideration. Its function, together with the vector scalar (dot) product is to select the component of the vector $\vec{F}$ that is normal to the surface S for integration. The surface integral is occasionally written as $\int_s F_n\, d\sigma$ where F_n means "normal component of $\vec{F}$ ". The direction of the unit normal $\hat{n}$, and therefore also F_n, will generally vary from point to point on an arbitary curved surface. Only for a plane does $\hat{n}$ have the same direction everywhere.

The volume integral contains the term $\nabla \cdot \vec{F}$ which is known as the "divergence of $\vec{F}$". The ∇ ("del") in this expression is the symbol for a vector differential operator written, in rectangular coordinates, as

$$\nabla = \hat{i}\,\frac{\partial}{\partial x} + \hat{j}\,\frac{\partial}{\partial y} + \hat{k}\,\frac{\partial}{\partial z} \qquad (2)$$

where $\hat{i}$, $\hat{j}$, and $\hat{k}$ are unit vectors associated respectively with the x, y, and z directions. These vectors obey the relations $\hat{i} \cdot \hat{j} = 0$ $\hat{i} \cdot \hat{i} = 1$, etc. and $|\hat{i}| = |\hat{j}| = |\hat{k}| = 1$. In other coordinate systems (viz. cylindrical polar, spherical, etc.) the form of the "∇" operator is always a vector differential operator. The result of taking the scalar vector (dot) product of ∇ and any vector is scalar. In rectangular coordinates, with $\vec{F}$ (x,y,z),

$$\nabla \cdot \vec{F} = \frac{\partial F_x}{\partial x} + \frac{\partial F_y}{\partial y} + \frac{\partial F_z}{\partial z} \tag{3}$$

Note that if $\vec{F}(x) = F_x\hat{i}$ only, then $\nabla \cdot \vec{F} = \frac{\partial F_x}{\partial x}$, which is just the usual derivative of $\vec{F}$ with respect to x.

The mathematical statement represented by Eq. (1) tells us,in words, that the integral of the normal component of the vector $\vec{F}$ over some closed surface is equal to the integral of the "three-dimensional derivative" $\nabla \cdot \vec{F}$, throughout a volume of space which is bounded by the surface S referred to in the surface integral. To introduce some physical significance to this argument, let us recall Gauss' Law, formulated in electrostatics as a method of calculating the electric field strength $\vec{E}$ when the spatial distribution of charge was known. Gauss' Law is usually written, in SI units, as

$$\int_S \vec{E} \cdot \hat{n} \, d\sigma = \frac{1}{\varepsilon_0} q_{enclosed} \tag{4}$$

where $q_{enclosed}$ is the electrostatic charge enclosed in the volume surrounded by the surface S, the so-called Gaussian surface. For a three dimensional problem the charge can be distributed spatially inside some volume V. Representing this volume charge distribution by $\rho(x,y,z)$ [units of coul/m^3], we can write

$$q_{enclosed} = \int_V \rho(x,y,z) \, dV \tag{5}$$

Combining Eqs. (4.95) and (4.96) gives

$$\int_S \hat{E} \cdot \hat{n} \, d\sigma = \frac{1}{\varepsilon_0} \int_V \rho(x,y,z) \, dV \tag{6}$$

Eq. (6) is known as the integral form of Gauss' Law. It follows from this and the divergence theorem (Eq.(1)) that an alternative statement, the differential form, is represented by $\nabla \cdot \vec{E} = \rho/\varepsilon_0$. Thus, we see that Gauss' Law can be formulated in either integral or differential form and that the connection is made through the mathematical statement relating surface and volume integrals, i.e. the divergence theorem.

For our purposes in deriving Kirchhoff's formula, it is necessary to rework Eq. (1) to incorporate the amplitude functions for incident and diffracted waves. Since the divergence theorem holds for any vector we will choose a form for $\vec{F}$ which will suit our objective. We will take $\vec{F} = u\nabla v - v\nabla u$ where u and v are scalar functions of (x,y,z); the vector nature is contained in the ∇ operator (see Eq. (2)). We make the assumption that u and v are wave functions satisfying

$$\nabla^2 u + k^2 u = 0 \tag{7}$$

and

$$\nabla^2 v + k^2 v = 0 \tag{8}$$

respectively. Here $\nabla^2 \equiv \nabla \cdot \nabla \frac{\partial^2}{\partial x^2} + \frac{\partial^2}{\partial y^2} + \frac{\partial^2}{\partial z^2}$ and $k = 2\pi/\lambda$ as usual. Eqs. (7) and (8) are known as wave equations. As an example, consider the one-dimensional case where there is u(x) only. Then Eq. (7) reduces to

$$\frac{\partial^2 u}{\partial x^2} + ku = 0 \tag{9}$$

This is the same equation we used earlier for one component of the electrical vector $\vec{E}$. It is easy to show that $u(x) = A \cos kx$ is a solution to Eq. (9) , as is $A \sin kx$ or $Ae^{\pm ikx}$. The three dimensional case is merely an extension of this argument. The need for Eqs. (7) and (8) will become apparent soon.

If we substitute $\vec{E} = u\nabla v - v\nabla u$ into Eq. (1) the left hand side becomes

$$\text{LHS} = \int_S (u\nabla v - v\nabla u) \cdot \hat{n}\, d\sigma = \int_S (u\nabla_n v - v\nabla_n u)\, d\sigma \tag{10}$$

where $\nabla_n u$ means normal component of gradient of u. For example, if the surface S is a sphere, when $\hat{n}$ is the unit vector in the radial direction (see Fig. 1) then

$$\nabla_n u = \frac{\partial u}{\partial r}$$

Or, if the surface is an infinite plane perpendicular to the x-axis, then $\hat{n} = \hat{i}$ and

$$\nabla_n u = \frac{\partial u}{\partial x}$$

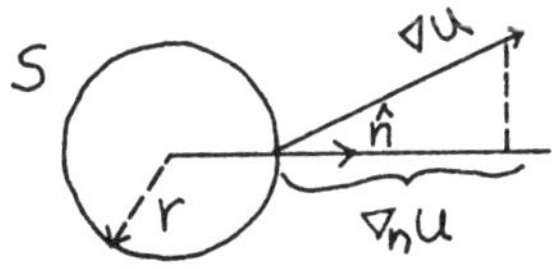

Fig. 1 Direction of the normal $\hat{n}$ and the gradient of u at a spherical surface. $\nabla_n u$ is the projection of ∇u onto the normal.

See Fig. 2 Consider the right hand side of Eq. (1)

$$\text{RHS} = \int_V \nabla \cdot (u\nabla v - v\nabla u)\, dv \tag{11}$$

Now

$$\nabla \cdot (u\nabla v) = \nabla u \cdot \nabla v + u\nabla^2 v \tag{12}$$

by a vector identity which is the analog of the derivative of a product for vector differentials. Using Eqs. (7) and (8) we now get

$$\begin{aligned} \text{RHS} &= \int_V [\nabla u \cdot \nabla v + u\nabla^2 v - (\nabla v \cdot \nabla u + v\nabla^2 u)\ ]\, dv \\ &= \int_V [u\nabla^2 v - v\nabla^2 u]\, dv = \int_V [u(-k^2 v) - v(-k^2 u)]\, dv = 0 \end{aligned} \tag{13}$$

That is, for our choice of u and v, the volume integral vanishes and we now have the result that

$$\int_S (u\nabla_n v - v\nabla_n u)\, d\sigma = 0 \tag{14}$$

Up to this point, u and v have not been specified except that they are scalar functions of position, satisfying the wave equation, and labelled with rectangular coordinates for simplicity. To obtain Kirchhoff's formula, u and v will be taken as the incident and diffracted wave amplitude, respectively. Since we will be interested in Fresnel conditions, represented by

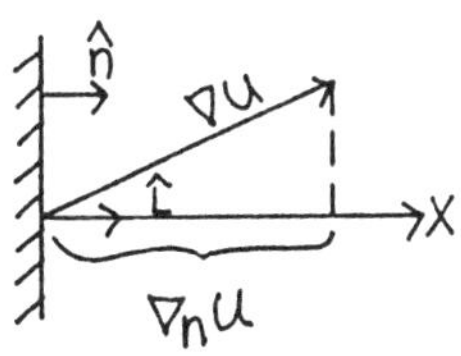

Fig. 2. Direction of the normal $\hat{n}$ and the gradient of u at a plane surface. $\nabla_n u$ is the projection of ∇u onto the normal.

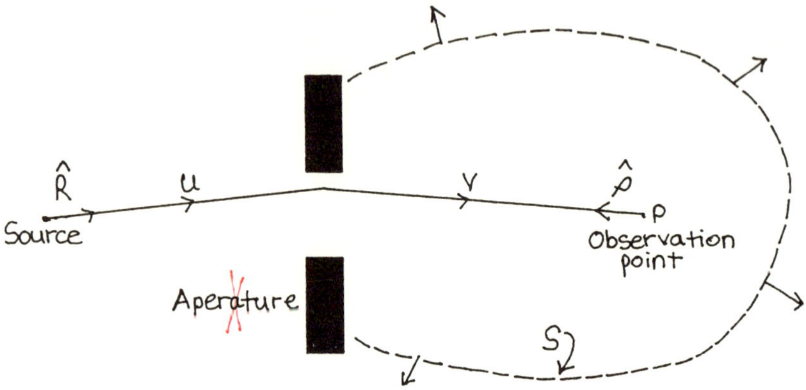

Fig. 3. Source point, observation point and integration surface for integration of Eq. (14).

spherical waves, it is clear that rectangular coordinates will not be the most appropriate. We can visualize our problem schematically in Fig. 3. The incident spherical wave $u = e^{ikR}/R$ excites the secondary Huygens' wave $v = e^{ik\rho}/\rho$. We now must consider what the surface S, in Eq. (14), must be to yield the desired result. Since we are interested in obtaining the amplitude at the observation point P, our surface must include this arbitrary point. Therefore, we take a surface composed of the aperture screen and a curved surface which a) includes P and closes onto the aperture screen and b) is large enough so that the "boundary effects" can be negligibly small. For convenience, this boundary is extended to ∞. The actual shape of S is irrelevant. We assume that u and v vanish on the aperture screen and at ∞. Before we substitute for u and v into Eq. (14), we notice that both scalar functions have singularities, i.e.

$$\lim_{R\to 0} u = \infty \text{ and } \lim_{\rho\to 0} v = \infty$$

Clearly, we must do something to avoid these singularities, since our goal is to obtain an expression for a well-behaved, physical quantity. The usual method of avoiding such a singularity is to surround it with an arbitrarily small surface S' which serves to exclude the singular point from the volume of interest. Then due consideration must be given to the effect of introducing S' on the quantity to be calculated. In Fig. 4 the volume of interest is shaded. Point P is surrounded by a spherical surface S' of radius ρ.

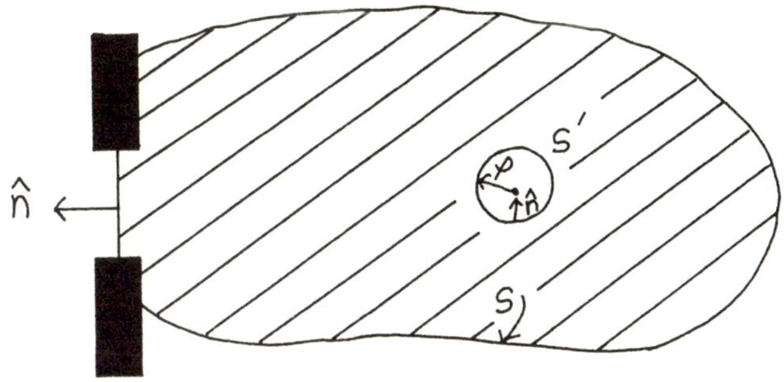

Fig. 4. Integration surfaces as in Fig. 3 The integration over the opening yields the incident light. A small sphere containing the observation point is indicated.

We now rewrite Eq. (14), taking into account both surfaces, as

$$\int_S (u\nabla_n v - v\nabla_n u)\, d\sigma + \int_{S'} (u\nabla_n v - v\nabla_n u)\, d\sigma' = 0 \tag{15}$$

Substituting $u = e^{ikR}/R$ and $v = e^{ik\rho}/\rho$, and evaluating each integral separately, we get first

$$\int_{S'} v\nabla_n u\, d\sigma' = \int_{S'} \frac{e^{ik\rho}}{\rho}\, \nabla_n \frac{e^{ikR}}{R}\, d\sigma' \tag{16}$$

Here since the source of u is far from P, we may treat $u \simeq$ const on S' so that $\nabla_n u = 0$ and

$$\int_{S'} v\nabla_n u\, d\sigma = 0 \tag{17}$$

Next consider

$$\int_{S'} u\nabla_n v\, d\sigma' = u\int_{S'} \nabla_n \frac{e^{ik\rho}}{\rho}\, d\sigma' \tag{18}$$

where u is, as before, the constant value of u on S'. To evaluate the gradient ∇_n on a surface of area $4\pi\rho^2$, we first define a new vector $\hat{\rho}$ which points radially outward from P. See Fig. 5. Then

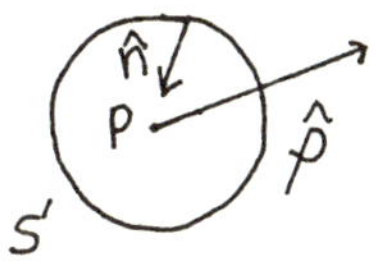

Fig. 5. Direction of $\hat{\rho}$ and $\hat{n}$ at the surface around the observation point P.

$$\nabla_n \frac{e^{ik\rho}}{\rho} = (\nabla \frac{e^{ik\rho}}{\rho}) \cdot \hat{n} = (\hat{\rho} \frac{\partial}{\partial\rho} \frac{e^{ik\rho}}{\rho}) \cdot \hat{n} = (\hat{\rho} \cdot \hat{n}) [\frac{ik}{\rho} - \frac{1}{\rho^2}] e^{ik\rho} \tag{19}$$

Then with $d\sigma' = \rho^2 \sin\theta' \, d\theta' \, d\psi'$, we get for Eq. (12.18)

$$u \int_{s'} \nabla_n \frac{e^{ik\rho}}{\rho} d\sigma' = u \int_{s'} (\hat{\rho} \cdot \hat{n})(ik\rho - 1) \frac{e^{ik\rho}}{\rho^2} \rho^2 \sin\theta' \, d\theta' \, d\psi' \tag{20}$$

$$= u \, (\hat{\rho} \cdot \hat{n}) \, [ik\rho e^{ik\rho} \int_{s'} \sin\theta' \, d\theta' \, d\psi' - e^{ik\rho} \int_{s'} \sin\theta' \, d\theta' \, d\psi']$$

Now since both $\hat{n}$ and $\hat{\rho}$ are oppositely directed on S', $(\hat{\rho} \cdot \hat{n}) = \cos\pi = -1$. Also, we require u to be evaluated at the observation point P which is obtained from the value on S' by letting $\rho \to 0$, that is

$$\lim_{\rho\to 0} [u \int_{s'} \nabla_n \frac{e^{ik\rho}}{\rho} d\sigma'] = u_P \lim_{\rho\to 0} [-ik\rho \int_{s'} e^{ik\rho} \sin\theta' \, d\theta' \, d\psi' + \int_{s'} e^{ik\rho} \sin\theta' \, d\theta' \, d\psi'] = 4\pi u_P \tag{21}$$

Therefore, we get for the integral over S',

$$\int_{s'} (u\nabla_n v - v\nabla_n u) \, d\sigma' = 4\pi u_P \tag{22}$$

We now consider the first term in Eq. (15), i.e. the integral over S,

$$\int_{s} (u\nabla_n v - v\nabla_n u) \, d\sigma = \int_{s} [\frac{e^{ikR}}{R} \nabla \frac{e^{ik\rho}}{\rho} - \frac{e^{ik\rho}}{\rho} \nabla \frac{e^{ikR}}{R}] \cdot \hat{n} \, d\sigma \tag{23}$$

Let R be a unit vector along R pointing from the source to the aperture, Then in the first integral

$$\nabla_n u = \nabla u \cdot \hat{n} = [R \frac{\partial}{\partial R} \frac{e^{ikR}}{R}] \cdot \hat{n} = (\hat{R} \cdot \hat{n}) [\frac{ik}{R} - \frac{1}{R^2}] e^{ikR} \tag{24}$$

and

$$\int_{s} v\nabla_n u \, d\sigma = \int_{s} \frac{e^{ik\rho}}{\rho} (\hat{R} \cdot \hat{n}) [\frac{ik}{R} - \frac{1}{R^2}] e^{ikR} \, d\sigma \tag{25}$$

Since the source is far from the aperture, R is large so that the R^{-2} term can be neglected. Also, we write $(\hat{R} \cdot \hat{r}) = \cos(\hat{R}, \hat{n})$ for convenience. Eq. (23) then becomes

$$\int_s v\nabla_n u \, d\sigma = \int_s \frac{e^{ik(R+\rho)}}{R\rho} \, ik \cos(\hat{R}, \hat{n}) \, d\sigma \tag{26}$$

The second integral in Eq (23) is obtained by similar considerations and is found to be

$$\int_s u\nabla_n v \, d\sigma = \int_s \frac{e^{ik(R+\rho)}}{R\rho} \, ik \cos(\hat{\rho}, \hat{n}) \, d\sigma \tag{27}$$

Finally, combining Eqs.(23),(26), and (27) gives

$$4\pi u_p = \int_s \frac{e^{ik(R+\rho)}}{R\rho} \, ik[\cos(\hat{R},\hat{n}) - \cos(\hat{\rho},\hat{n})] \, d\sigma$$

or

$$u_p = \frac{ik}{4\pi} \int_s \frac{e^{ikR}}{R} [\cos(\hat{R},\hat{n}) - \cos(\hat{\rho},\hat{n})] \frac{e^{ik\rho}}{\rho} \, d\sigma \tag{28}$$

Eq. (28) is known as the Fresnel-Kirchhoff Integral.

The derivation presented here, is still only an approximation. It does not clearly state the boundary values for v and dv/dn. The mathematical difficulties connected with this problem are discussed in Sommerfeld's text on theoretical optics.*

---

4 Maxwell's Theory

1. *Reflection on an optically denser medium.*
   a. Plot $r_{E_\perp}$, $r_{E_\parallel}$, $t_{E_\perp}$ and $t_{E_\parallel}$ for $N_2 = 1.5$ and $n_1 = 1$, where the angle of incidence θ is equal to 20°, 40°, 60°, and 80°.
   b. Find the angle for which $r_{E_\parallel}$ is zero (read off your graph). Compare this with $\tan\theta_1 = N_2/n_1$.

* A. Sommerfeld - Vorlesungen uber Theoretische Physik, Vol. IV, Optik- Dietrich'sche Verlagsbuchhandlung Wiesbaden 1950.

Solution

(a)

θ	$r_{E\perp}$	$r_{E\parallel}$	$t_{E\perp}$	$t_{E\parallel}$
20°	-.217	-.183	.783	.789
40°	-.277	-.120	.722	.746
60°	-.420	.042	.579	.638
80°	-.733	.487	.266	.342

(b) $\theta_{1B} = 56.3°$

Plots of data

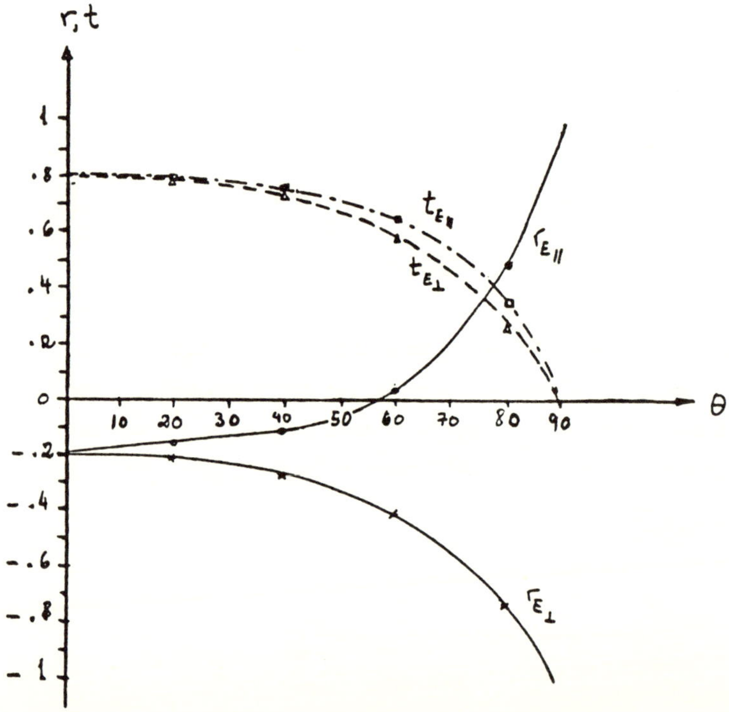

2. *Dielectric plate and Brewster's angle.* Unpolarized light is incident on a dielectric plate ($N_2 = 1.45$) at Brewster's angle.
   a. Calculate $r_\parallel^2$, $r_\perp^2$, $\alpha t_\parallel^2$ and $\alpha t_\perp^2$, where

$$\alpha = \frac{N_2 \cos\theta''}{n_1 \cos\theta}$$

   b. Show that $r_\perp^2 + \alpha t_\perp^2 = 1$ and $r_\parallel^2 + \alpha t_\parallel^2 = 1$.

   c. Give the percentage of the transmitted light of the incident unpolarized light.

   d. For a check, calculate the percentage for the reflected light.

Solution

(a) $r_\parallel{}^2 = 0$ $\qquad$ $\tan\theta = 1.450$ $\qquad$ $\theta = 55.408$

$$r_\perp = \frac{\cos\theta - N_2\sqrt{1 - \left(\frac{\sin\theta}{N_2}\right)^2}}{\cos\theta + N_2\sqrt{1 - \left(\frac{\sin\theta}{N_2}\right)^2}} = -.356$$

$$(r_\perp)^2 = .127$$

$$\alpha = 1.450\,\frac{\sqrt{1 - \frac{\sin^2\theta}{(1.450)^2}}}{\cos\theta} = 2.105$$

$$t_\parallel = \frac{2\cos\theta}{1.450\cos\theta + \sqrt{1 - \frac{\sin^2\theta}{(1.450)^2}}} = .689$$

$$t_\parallel{}^2 = .475 \qquad \alpha\, t_\parallel{}^2 = 1.000$$

$$t_\perp = \frac{2\cos\theta}{\cos\theta + 1.450\sqrt{1 - \frac{\sin^2\theta}{(1.450)^2}}} = .644$$

$$t_\perp{}^2 = .415 \qquad \alpha\, t_\perp{}^2 = .874$$

(b) $r_\perp{}^2 + \alpha\, t_\perp{}^2 = 1.001$

$r_\parallel{}^2 + \alpha\, t_\parallel{}^2 = 1.000$

(c) Incident 1 + 1 = 2

Transmitted 1 + .874 = 1.874

Percentage transmitted $\frac{1.874}{2}$ 100 = 93.7

(d) Reflected 0 + .127 = .127

Percentage reflected $\frac{.127}{2} \cdot 100 = 6.35$

Check 6.35% + 93.7% = 100.05%

3. *Transmission close to Brewster's angle for glass.* In section 3F, we gave Brewster's angle for glass ($n = 1.5$) as 56.3°.
   a. Calculate $r_\parallel$ for $\theta = 56°$, and calculate $r_\parallel^2$.
   b. Calculate the transmitted light $I_\parallel = 1 - r_\parallel^2$ and compare with 1, which is the value for the exact Brewster's angle. (Use five digits after decimal point.)
   c. Calculate $I_\parallel^{10}$, $I_\parallel^{100}$, $I_\parallel^{10,000}$, where $I_\parallel$ is the transmitted intensity. (I^n is equivalent to n times transversal of the interface).

Solution

$n_2 = 1.50000$

(a) $$r_{\parallel} = \frac{-n_2 \cos 56 + \sqrt{1 - (\frac{\sin 56}{n_2})^2}}{n_2 \cos 56 + \sqrt{1 - (\frac{\sin 56}{n_2})^2}} = -.00323$$

$$r_{\parallel}^2 = .00001$$

(b) $I_{\parallel} = 1 - r_{\parallel}^2 = .99999$

(c) $I_{\parallel}^{10} = .99990$, $I_{\parallel}^{100} = .99900$, $I_{\parallel}^{10000} = .90484$

4. *Amplitude reflection coefficient.* In Chapter 2, we saw that the amplitude reflection coefficient r for reflection on the optical denser medium and the amplitude reflection coefficient r' for reflection on the optical less dense medium are related by $r = -r'$. This was obtained from Stoke's consideration. Show that it also follows from Fresnel's formulas.

Solution

For r_s from medium n_1 to medium n_2, $n_1 < n_2$

$$r_s = \frac{n_1 \cos \theta - n_2 \cos \theta''}{n_1 \cos \theta + n_2 \cos \theta''}$$

For r_s' from medium n_2 to medium n_1, $n_2 > n_1$

$$r_s' = \frac{n_2 \cos \theta'' - n_1 \cos \theta}{n_2 \cos \theta'' + n_1 \cos \theta}$$

We have $r_s = -r_s'$

$$r_p = -\frac{(n_2 \cos \theta - n_1 \cos \theta'')}{n_2 \cos \theta + n_1 \cos \theta''}$$

$$r_p' = \frac{-(n_1 \cos \theta'' - n_2 \cos \theta)}{n_1 \cos \theta'' + n_2 \cos \theta}$$

It follows $r_p = -r_p'$

5. *Reflection on the optically less dense medium.*
   a. Plot $r_{E_\perp}$, $r_{E_\parallel}$, $t_{E_\perp}$, and $t_{E_\parallel}$ for $n_2 = 1.0$ and $N_1 = 1.5$, where $\theta_2 = 20°$, $40°$, $60°$, and $80°$.
   b. Find the angle for which $r_{E_\parallel}$ is zero (read off your graph). Compare this with $\tan \theta_2 = n_2/N_1$.
   c. What happens in the range where you get imaginary numbers?

Solution

(a)

θ_2	$r_{E\perp}$	$r_{E\parallel}$	$t_{E\perp}$	$t_{E\parallel}$
20°	.243	.053	1.243	1.27
40°	.625	-.316	1.62	1.96
60°				
80°				

(b)

$\theta_{2B} = 33.69°$

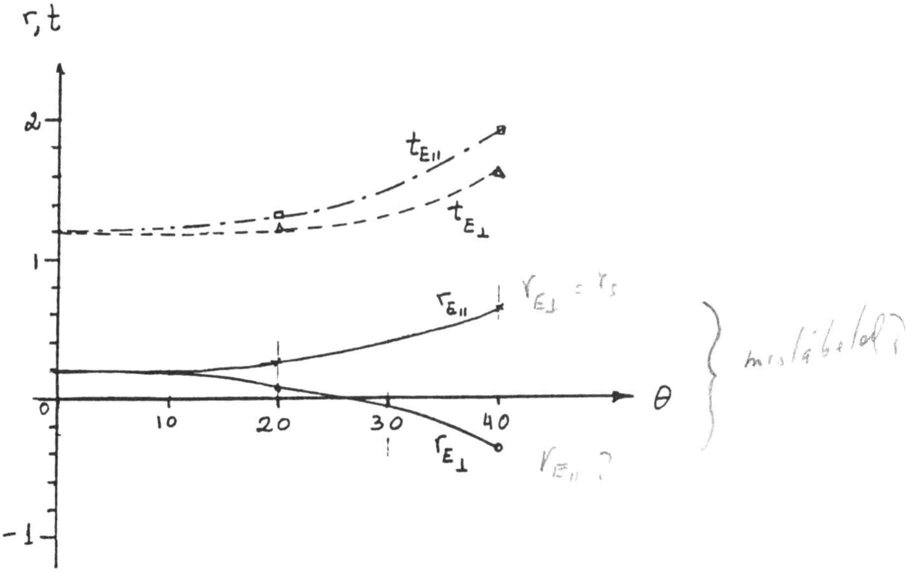

(c) Total internal reflection for $\theta_2 = 60°$, and $\theta_2 = 80°$

6. *Brewster's angle at different media.* When light is incident at Brewster's angle on a plane parallel plate of glass, such as used for a laser window, it hits the glass-air interface also at Brewster's angle.
   a. Show that the law of refraction is obeyed by calculation with formulas.
   b. Show that the law of refraction is obeyed by numerically calculating Brewster's angle and inserting it into the law of refraction.

Solution

(a) We have $\theta_1 + \theta_2 = 90°$, if θ_1 is Brewster angle

From $\tan\theta_1 = \dfrac{n_2}{n_1}$

$$n_1 \sin\theta_1 = n_2 \cos\theta_1 = n_2 \cos(90-\theta_2) = n_2 \sin\theta_2$$

(b) $\theta_{B1} = 56.3$, $\theta_{B2} = 33.7$ $1. \sin 56.3 = 1.5 \sin 33.7$

7. *Degree of polarization.*

a. The degree of polarization q may be defined as

$$q = \frac{I_p - I_s}{I_p + I_s}$$

where I_p and I_s are the transmiited intensities for the p and s polarization, respectively. If we define

$$a = \frac{n_1 \cos\theta'' + n_2 \cos\theta}{n_1 \cos\theta + n_2 \cos\theta''}$$

show that q can be written as

$$q = \frac{1 - a^2}{1 + a^2}$$

b. Calculate q for the angles of incidence 20°, 40°, 60°, and 80°, ($n_1 = 1, n_2 = 1.5$) and make a plot.

Solution

(a) $$\frac{t_s}{t_p} = \frac{\dfrac{2 n_1 \cos\theta}{n_1 \cos\theta + n_2 \cos\theta''}}{\dfrac{2 n_1 \cos\theta}{n_2 \cos\theta + n_1 \cos\theta''}} = \frac{n_2 \cos\theta + n_1 \cos\theta''}{n_1 \cos\theta + n_2 \cos\theta''}$$

$$\frac{I_s}{I_p} = \left(\frac{t_s}{t_p}\right)^2 = a^2 \qquad q = \frac{I_p - I_s}{I_p + I_s} = \frac{1 - \dfrac{I_s}{I_p}}{1 + \dfrac{I_s}{I_p}} = \frac{1 - a^2}{1 + a^2}$$

(b)

θ	q
20	7.1×10^{-3}
40	.03
60	.1
80	.25

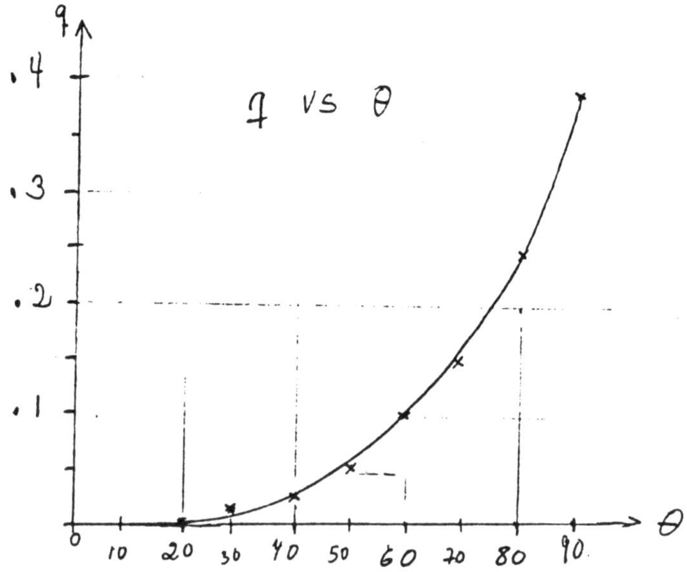

8. *Partly polarized light, parallel component.* Consider light passing through a plate of glass ($n = 1.5$) under angles of incidence of 20°, 40°, 60°, and 80°.
   a. Calculate the angle of refraction θ''.
   b. The angles θ'' are the angles of incidence when the ray hits the second (inner) surface of the glass plate. Using Fresnel's formula, calculate the power $1 - r_p^2 = \alpha t_p^2$ transmitted through the *first* interface for the component parallel to the plane of incidence.
   c. Calculate the transmitted intensity through the second interface for the parallel component.
   d. Combine (b) and (c) and calculate the total intensity P_p transmitted for the parallel component.

Solution

(a)

θ_i	θ''
20°	13.2°
40°	25.4°
60°	35.3°
80°	41.0°

(b)

θ_i	$(1-r_p^2)_1$
20°	.97
40°	.99
60°	1.00
80°	.76

(c)

θ_i	$(1-r_p^2)_2$
13.2	.97
25.4	.99
35.3	1.00
41.0	.76

(d)

θ_i	$(1-r_p^2)_1(1-r_p^2)_2$
20°	.94
40°	.98
60°	1.00
80°	.58

9. *Partly polarized light, perpendicular component.* Calculate, in the same way as in problem 8, the transmitted intensity for the perpendicular component.
   a. Use the angles of refraction as calculated in problem 8a.
   b. Calculate $(1 - r_s^2)_1$, transmitted through the first interface.
   c. Calculate $(1 - r_s^2)_2$, transmitted through the second interface.
   d. Calculate $(1 - r_s^2)_1 \cdot (1 - r_s^2)_2$ and plot it on the same graph with the data of $(1 - r_p^2)_1 \cdot (1 - r_p^2)_2$ from problem 8d.

Solution

(a)

θ_i	θ_t
20°	13.2°
40°	25.4°
60°	35.3°
80°	41.0°

(b)

θ_i	$(1-r_s^2)_1$
20°	.95
40°	.92
60°	.82
80°	.46

(c)

θ_i	$(1-r_s^2)_2$
13.2°	.95
25.4°	.92
35.3°	.82
41.0°	.46

(d)

θ_i	$(1-r_s^2)_1(1-r_s^2)_2$
20°	.90
40°	.85
60°	.67
80°	.21

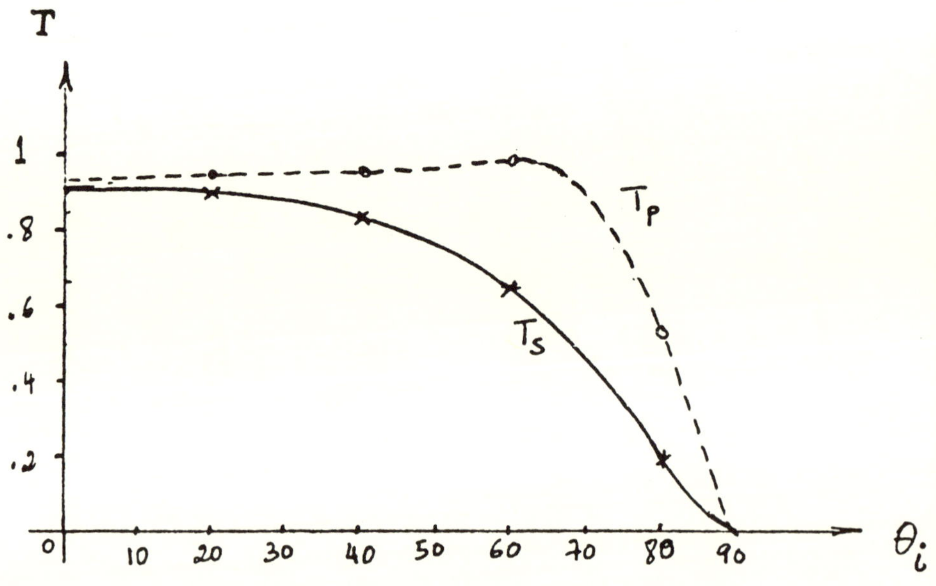

10. *Critical angle for glass.* We want to couple a light beam into a plane parallel plate using a prism, as shown.

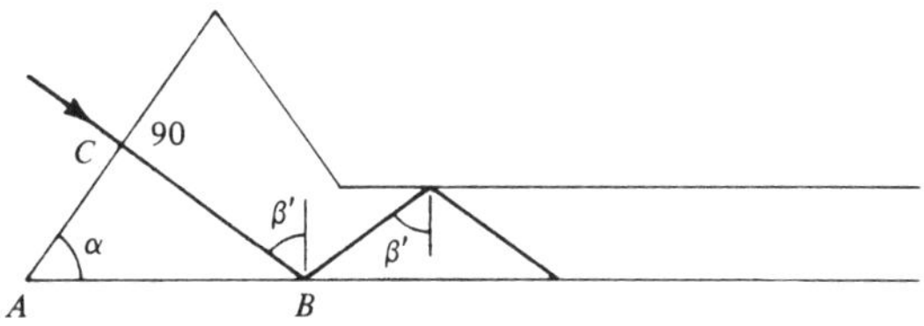

The light should be at normal incidence on the prism.

a. What is the relation between α and β'?

b. How should we choose α so that the waves get internally reflected? The refractive index is $n = 1.5$ (glass).

Solution

(a) It is from the trangle A B C

$$180 = 90 + \alpha + (90-\beta')$$

It follows $\alpha = \beta'$

(b) Critical angle $\sin\theta_c = \frac{n_2}{N_1} = \frac{1}{1.5}$

$$\theta_c = 41.8°$$

$41.8° \le \alpha \le 90°$

11. *Phase calculation for total reflection (sample calculation).* Calculate the phase angle of the reflected waves $r_{E_\parallel}$ and $r_{E_\perp}$ for reflection of the less dense medium. Take $N_1 = 1.5$, $n_2 = 1$, and find the phase angles $\delta_\perp$ and $\delta_\parallel$ for $\theta_c < 0 < 90°$; make a plot.

Solution

We have $r_\perp = \dfrac{\cos\theta - \sqrt{(\frac{n_2}{N_1})^2 - \sin^2\theta}}{\cos\theta + \sqrt{(\frac{n_2}{N_1})^2 - \sin^2\theta}}$

For $\theta>\theta_c$ we have $(\frac{n_2}{N_1})^2 < \sin^2\theta$

We introduce $\sqrt{-1} = i$ and have

$$r_\perp = \frac{\cos\theta - i\sqrt{\sin^2\theta - (\frac{n_2}{N_1})^2}}{\cos\theta + i\sqrt{\sin^2\theta - (\frac{n_2}{N_1})^2}}$$

We set $Z = \cos\theta - i\sqrt{\sin^2\theta - (\frac{n_2}{N_1})^2}$

and have $r_\perp = \frac{Z}{Z^*}$. We set $Z = Ae^{i\alpha}$ then $r_\perp = e^{i2\alpha}$ and this is equal to $e^{i\delta_\perp}$. $\delta_\perp$ is what we want to determine. We get

α from $\tan\alpha = \frac{\text{Im}(Z)}{\text{Re}(Z)}$

$$\tan\alpha = \frac{-\sqrt{\sin^2\theta - (\frac{n_2}{N_1})^2}}{\cos\theta}$$

$$\text{and } \delta_\perp = -2 \text{ actan} \frac{\sqrt{\sin^2\theta - (\frac{n_2}{N_1})^2}}{\cos\theta}$$

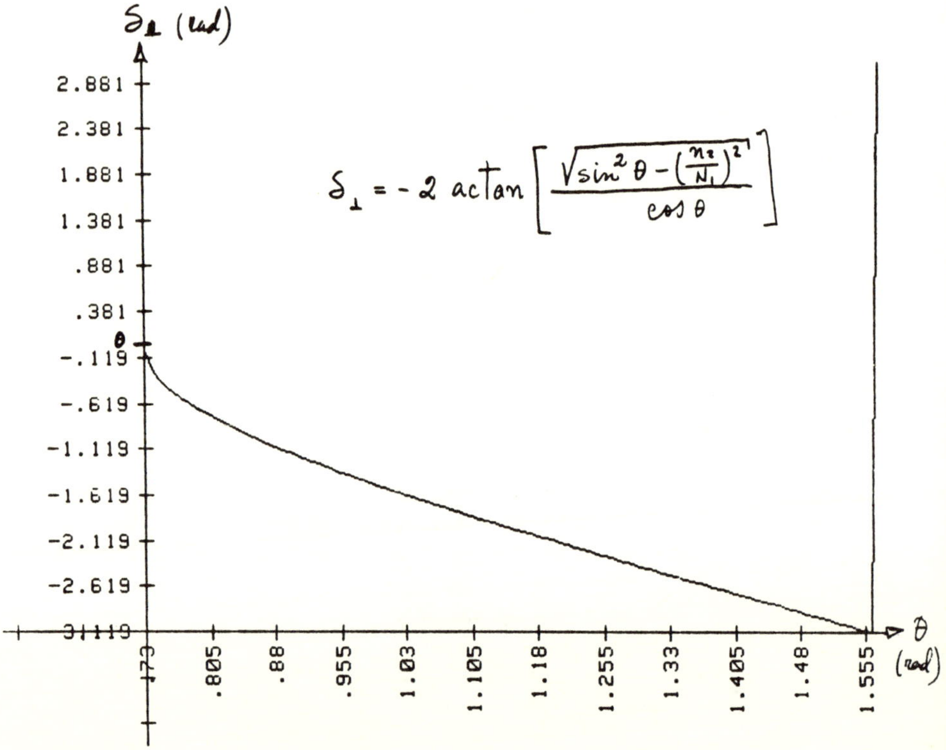

$$\text{For } r_\parallel \text{ we have } \frac{-\frac{n_2}{N_1}\cos\theta + \sqrt{1 - (\frac{N_1}{n_2})^2 \sin^2\theta}}{\frac{n_2}{N_1}\cos\theta + \sqrt{1 - (\frac{N_1}{n_2})^2 \sin^2\theta}}$$

For $\theta > \theta$ we have $(\frac{N_1}{n_2})^2 \sin^2\theta > 1$ and write

$$r_{\parallel} = \frac{-\frac{n_2}{N_1}\cos\theta + i\sqrt{(\frac{N_1}{n_2})^2\sin^2\theta - 1}}{\frac{n_2}{N_1}\cos\theta + i\sqrt{(\frac{N_1}{n_2})^2\sin^2\theta - 1}}$$

Calling $-\frac{n_2}{N_1}\cos\theta + i\sqrt{(\frac{N_1}{n_2})^2\sin^2\theta - 1} = Z_1$

We have for $r_{\parallel} = -\frac{Z_1}{Z_1^*} = e^{i\delta_{\parallel}}$

Using the same steps of calculations we used above for $\delta_{\perp}$, we obtain

$$\delta_{\parallel} = 180 - 2\ \text{actan}\ \left[\frac{\sqrt{(\frac{N_1}{n_2})^2\sin^2\theta - 1}}{\frac{n_2}{N_1}\cos\theta}\right]$$

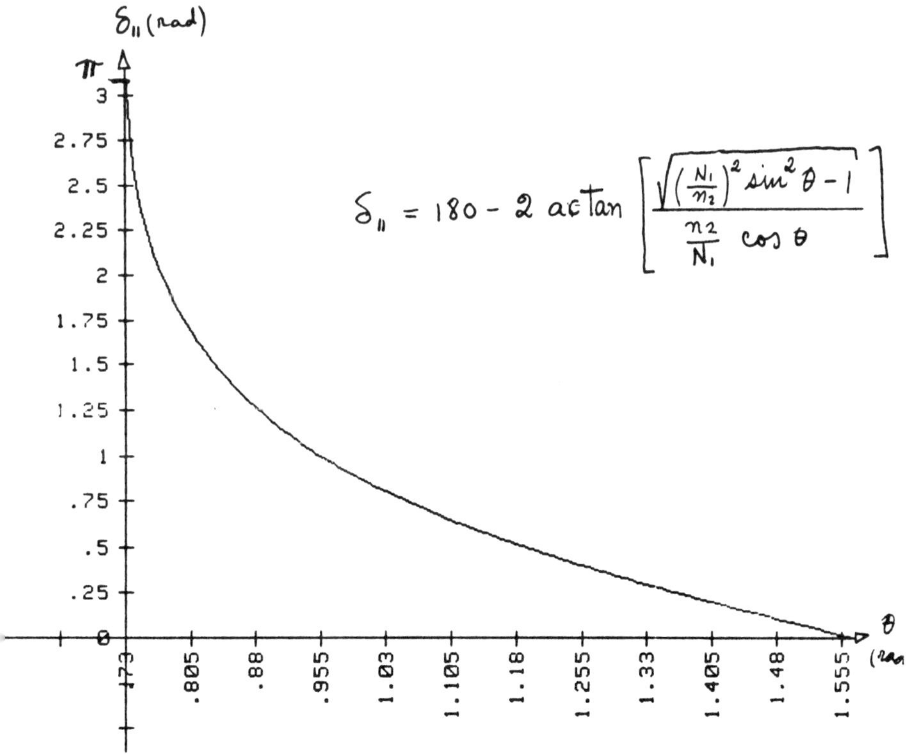

12. *Attenuation at total reflection in the less dense medium.* The function

$$y = Ae^{-|Y|(2\pi/\lambda)\sqrt{(n\sin\theta)^2-1}}$$

is the penetration depth at angle $\theta > \theta_c$ for a wave incident on a glass-air interface. Plot y as a function of $|Y|$. To do this, take $y = A = 1$ for $|Y| = 0$, $\lambda = 0.5\cdot 10^{-4}$ cm, and $n = 1.5$ for glass. Plot y for $\theta = \theta_c + 2°$ for $|Y| = 0.1\lambda, 0.3\lambda, 0.5\lambda, 0.7\lambda$, and 0.9λ.

Solution

$\theta_c = \sin^{-1}\left(\frac{1}{1.5}\right) = 41.8°$ $\qquad\theta = 41.8 + 2 = 43.8°$

$y = e^{-2\pi(.1)\sqrt{(1.5\sin 43.8)^2-1}} = e^{-3.142\cdot 0.279(.1)}$

$e^{-1.758\cdot(.1)}$

Y	X	$y=e^{-X}$
.1	.175	.839
.3	.524	.590
.5	.875	.415
.7	1.225	.292
.9	1.575	.206

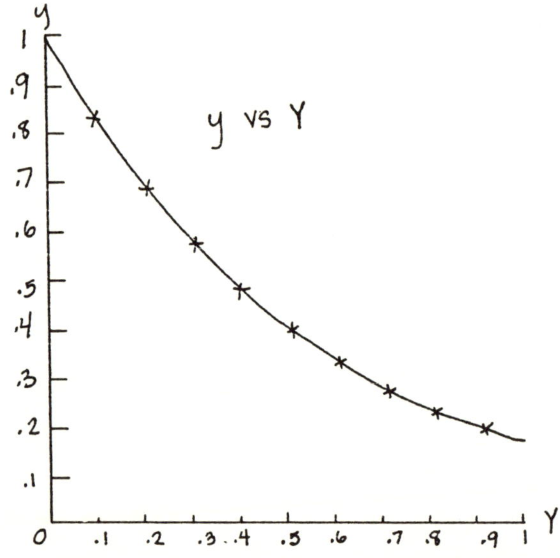

13. *Inverse matrix.* Find the inverse matrix of the following 2 × 2 matrix.

$$\begin{pmatrix} 1 & 1 \\ -n_2 & n_2 \end{pmatrix}$$

Solution

Inverse Matrix of matrix (a_{ij}):

co-factor of a_{ij} is $A_{ij} = (-1)^{i+j}M_{ij}$ where M_{ij} is the minor of the element a_{ij}.

Inverse matrix of (a_{ij}) is $\frac{(A_{ji})}{\text{Det.a}}$ where Det.a is the determinate of the matrix (a_{ij})

$a_{ij} = \begin{pmatrix} 1 & 1 \\ -n_2 & n_2 \end{pmatrix}$ $\qquad a_{ji} = \begin{pmatrix} 1 & -n_2 \\ 1 & n_2 \end{pmatrix}$

$$M_{ji} = \begin{pmatrix} n_2 & 1 \\ -n_2 & 1 \end{pmatrix} \qquad A_{ji} = \begin{pmatrix} n_2 & -1 \\ n_2 & 1 \end{pmatrix}$$

$$\text{Det.a} = 2n_2$$

$$(a_{ij})^{-1} = \begin{pmatrix} \frac{1}{2} & -\frac{1}{2n_2} \\ \frac{1}{2} & \frac{1}{2n_2} \end{pmatrix}$$

Proof: $$\begin{pmatrix} 1 & 1 \\ -n_2 & n_2 \end{pmatrix} \begin{pmatrix} \frac{1}{2} & -\frac{1}{2n_2} \\ \frac{1}{2} & \frac{1}{2n_2} \end{pmatrix} = \begin{pmatrix} 1 & 0 \\ 0 & 1 \end{pmatrix}$$

14. *Fabry-Perot filter: I.* Consider a device called a Fabry-Perot filter, which is composed of many layers of films, each with the same thickness $\lambda/4$ but with different indices of refraction. The films have either a high index n_H or a low index n_L. The corresponding matrices are

$$M_H = \begin{pmatrix} 0 & -\frac{i}{n_H} \\ -in_H & 0 \end{pmatrix} \qquad M_L = \begin{pmatrix} 0 & -\frac{i}{n_L} \\ -in_L & 0 \end{pmatrix}$$

Assume there is an alternating sequence of layers, symmetrically arranged with respect to the center and having an even number of layers on each side:

$$|n_H|n_L|n_H|n_L|\ldots|n_H|n_L|n_L|n_H|\ldots|n_L|n_H|n_L|n_H|$$
$$\uparrow$$
$$\text{Center}$$

The matrix representation of this device is

$$M_H M_L M_H M_L \ldots M_H M_L M_L M_H \ldots M_L M_H M_L M_H$$

Show that the product of all the matrices is the unit matrix

$$\begin{pmatrix} 1 & 0 \\ 0 & 1 \end{pmatrix}$$

(*Hint:* Start with the calculation of $M_L M_L$ in the middle.) The result tells us that for the particular wavelength $\lambda = 4d$, the film is completely transparent. Why do we need so many layers? (To eliminate the other wavelengths.)

<u>Solution</u>

Consider $$M_H\, M_L\, M_H \,\cdots\cdots\, M_H\, \underbrace{M_L\, M_L}_{\text{center}}\, M_H \,\cdots\cdots\, M_L\, M_H\, M_L$$

$$M_L\, M_L = \begin{pmatrix} 0 & -\frac{i}{n_L} \\ -in_L & \end{pmatrix} \begin{pmatrix} 0 & -\frac{i}{n_L} \\ -in_L & 0 \end{pmatrix} = \begin{pmatrix} -1 & 0 \\ 0 & -1 \end{pmatrix}$$

We get for

$$M_H\, M_L\, M_L\, M_H = -\begin{pmatrix} 0 & -\frac{i}{n_H} \\ -in_H & 0 \end{pmatrix} \begin{pmatrix} 1 & 0 \\ 0 & 1 \end{pmatrix} \begin{pmatrix} 0 & \frac{-i}{n_H} \\ -in_H & 0 \end{pmatrix} = \begin{pmatrix} 1 & 0 \\ 0 & 1 \end{pmatrix}$$

For even number of layers (on one side) the product is

$$\begin{pmatrix} 1 & 0 \\ 0 & 1 \end{pmatrix}$$

We have

$$\begin{pmatrix} E \\ B \end{pmatrix}_{\text{end}} = \begin{pmatrix} 1 & 0 \\ 0 & 1 \end{pmatrix} \begin{pmatrix} E \\ B \end{pmatrix}_{\text{begin}}$$

For $\lambda \neq 4d$ this is not the case

15. *Fabry-Perot filter: II.* Consider a filter made of alternating layers (a Fabry-Perot filter). Assume that there are N of each kind of layers. The sequence of the refractive indices is then

$$n = 1 \qquad \underbrace{|n_H|n_L|}_{N\text{ times}}n_H|\ldots|n_L| \qquad n = 1$$

Show that for the matrix product we get

$$\begin{pmatrix} \left(-\dfrac{n_L}{n_H}\right)^N & 0 \\ 0 & \left(-\dfrac{n_H}{n_L}\right)^N \end{pmatrix}$$

That is, we have

$$\begin{pmatrix} A_f \\ -A_f \end{pmatrix} = \begin{pmatrix} \left(-\dfrac{n_L}{n_H}\right)^N & 0 \\ 0 & \left(-\dfrac{n_H}{n_L}\right)^N \end{pmatrix} \begin{pmatrix} A_1 + A_1' \\ -A_1 + A_1' \end{pmatrix}$$

<u>Solution</u>

The product of M_H and M_L is

$$M_H\, M_L = \begin{pmatrix} 0 & -i/n_H \\ -in_H & 0 \end{pmatrix} \begin{pmatrix} 0 & -i/n_L \\ -in_L & 0 \end{pmatrix} = \begin{pmatrix} -n_L/n_H & 0 \\ 0 & -n_H/n_L \end{pmatrix}$$

And for $(M_H\, M_L)^2$

$$(M_H\ M_L)^2 = = \begin{pmatrix} -n_L/n_H & 0 \\ 0 & -n_H/n_L \end{pmatrix}\begin{pmatrix} -n_L/n_H & 0 \\ 0 & -n_H/n_L \end{pmatrix} = \begin{pmatrix} [-n_L/n_H]^2 & 0 \\ 0 & [-n_H/n_L]^2 \end{pmatrix}$$

For $(M_H\ M_L)^N$

$$(M_H\ M_L)^N = \begin{pmatrix} [-n_L/n_H]^N & 0 \\ 0 & [-n_H/n_L]^N \end{pmatrix}$$

Form $\begin{pmatrix} E \\ B \end{pmatrix}_{end} = \begin{pmatrix} \alpha & 0 \\ 0 & \beta \end{pmatrix}\begin{pmatrix} E \\ B \end{pmatrix}_{begin}$ $\quad \alpha = [-n_L/n_H]^N \quad \beta = [-n_H/n_L]^N$

we get

$$\begin{pmatrix} A_f \\ -A_f \end{pmatrix} = \begin{pmatrix} \alpha & 0 \\ 0 & \beta \end{pmatrix}\begin{pmatrix} A_1+A_1' \\ -A_1+A_1' \end{pmatrix}$$

16. *Fabry-Perot filter: III.*

a. Show that we obtain for the reflection and transmission coefficients

$$\frac{A_f}{A_1} = \frac{2}{[-(n_H/n_L)]^N + [-(n_L/n_H)]^N} \quad \text{and} \quad \frac{A_1'}{A_1} = \frac{[-(n_L/n_H)]^N - [-(n_H/n_L)]^N}{[-(n_L/n_H)]^N + [-(n_H/n_L)]^N}$$

b. Show that the reflected intensity is equal to

$$R = \left[\frac{(n_H/n_L)^{2N} - 1}{(n_H/n_L)^{2N} + 1}\right]^2$$

c. Show that if $n_H = 2.5$, $n_L = 1.5$, and $N = 20$, then the reflected intensity is $R = 0.99999999$.

Solution

(a) $A_N = \alpha\ (A_1+A_1')$ $\qquad -A_N = \beta(-A_1+A_1')$

$$\frac{A_N}{A_1} = \alpha + \frac{A_1'}{A_1}\ \alpha \qquad -\frac{A_N}{A_1} = -\beta+\beta\ \frac{A_1'}{A_1}$$

To solve $(A_N/A_1) - \alpha(A_1'/A_1) = \alpha$

$-A_N/A_1 - \beta\ A_1'/A_1 = -\beta$

get

$$\frac{A_N}{A_1} = \frac{\begin{vmatrix} \alpha & -\alpha \\ -\beta & -\beta \end{vmatrix}}{\begin{vmatrix} 1 & -\alpha \\ -1 & -\beta \end{vmatrix}} = \frac{-2\alpha\beta}{-\beta-\alpha} = \frac{2\alpha\beta}{\alpha+\beta} = \frac{2}{(-\frac{n_H}{n_L})^N + (-\frac{n_L}{n_H})^N}$$

And for $\frac{A_1'}{A_1}$ we get

$$\frac{A_1'}{A_1} = \frac{\begin{vmatrix} 1 & \alpha \\ -1 & -\beta \end{vmatrix}}{\begin{vmatrix} 1 & -\alpha \\ -1 & -\beta \end{vmatrix}} = \frac{-\alpha+\beta}{\alpha+\beta} = \frac{-(-\frac{n_L}{n_H})^N + (-\frac{n_H}{n_L})^N}{(-\frac{n_L}{n_H})^N + (-\frac{n_H}{n_L})^N}$$

(b) $$R = \left(\frac{A_1'}{A_1}\right) = \left(\frac{-\alpha+\beta}{\alpha+\beta}\right)^2 = \left(\frac{-1+\frac{\beta}{\alpha}}{1+\frac{\beta}{\alpha}}\right)^2 = \left[\frac{-1+(\frac{n_H}{n_L})^{2N}}{1+(\frac{n_H}{n_L})^{2N}}\right]^2$$

$$R = \left[\frac{(\frac{n_H}{n_L})^{2N} - 1}{(\frac{n_H}{n_L})^{2N} + 1}\right]^2$$

(c) $$R = \left[\frac{-1 + (2.5/1.5)^{40}}{1 + (2.5/1.5)^{40}}\right]^2 \qquad (2.5/1.5)^{40} = 7.48 \times 10^8$$

$$R = \left(\frac{-1 + 748083342.7}{1 + 748083342.7}\right)^2 = .9999999947$$

---

5 Electromagnetic Waves In Crystals

1. *Wavelength in crystal.* Unpolarized light of wavelength $\lambda_0 = 0.5 \times 10^{-4}$ cm is incident normally on a crystal quartz plate with $n_o = 1.544$ and $n_e = 1.553$. The optic axis is in the plane of the interface. Calculate the wavelength of the ordinary and extraordinary waves in the quartz plate.

Solution

$$\lambda_o = .5.10^{-4}\text{cm} \qquad n_o = 1.544 \qquad n_e = 1.553$$

$$\lambda_{no} = \frac{\lambda_o}{n_o} = .3238\text{x}10^{-4}\text{cm ordinary}$$

$$\lambda_{ne} = \frac{\lambda_o}{n_e} = .3220\text{x}10^{-4}\text{cm extraordinary}$$

2. *Ordinary and extraordinary rays in calcite.* Consider a calcite prism with $n_o = 1.658$ and $n_e = 1.486$. The prism has the shape shown in the figure, with apex angle A. The light is incident parallel from the left. The optic axis is perpendicular to the plane of the paper, indicated by dots.

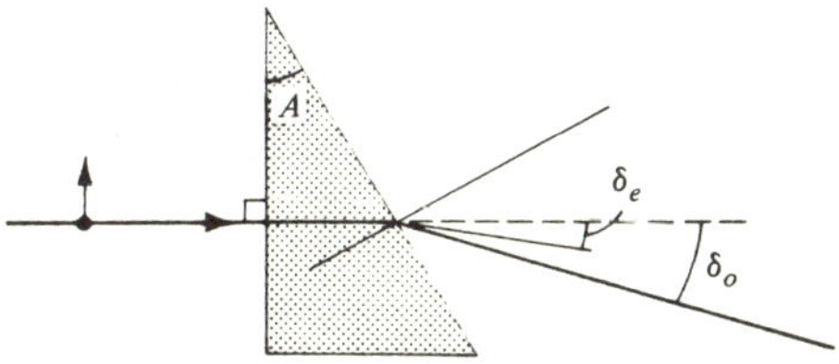

a. Calculate the angle of deviation δ_o for the ordinary ray. Do this for $A = 10°$, $20°$, and $30°$.
b. Calculate the angle of deviation δ_e for the extraordinary ray, again for $A = 10°$, $20°$, and $30°$.
c. Find $\Delta = \delta_o - \delta_e$ for the three angles.

Solution

for (a)

A	δ_o	δ_e	Δ
10	6.73°	4.95	1.78
20	14.55°	10.55	4.00
30	26.00	17.99	8.01

3. *Critical angle, calcite, and quartz.* Consider a plate of calcite with the optic axis perpendicular to the plane of the paper (indicated by dots in the accompanying figure). An unpolarized beam of light is entering the part of the plate that is shaped like a prism. The light enters the prism at normal incidence. (This relates α and β'.) The refractive indices are $n_o - 1.658$ and $n_o = 1.486$.

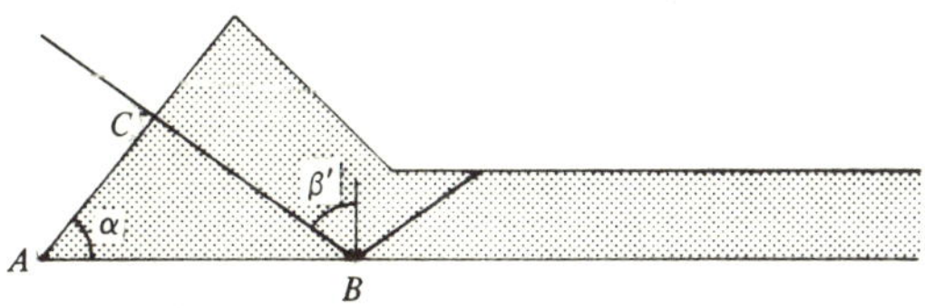

a. Calculate the critical angle β' for the ordinary and extraordinary rays.
b. If α is close to 40° (that is if the light is incident close to 40° at B), which ray will escape from the plate?
c. For quartz, $n_o = 1.544$ and $n_e = 1.553$. Calculate the critical angles.
d. Give a value of the angle α such that one ray stays in the plate and the other escapes. Which stays in the plate?

Solution

(a) $\theta_{co} = 37.09°$ $\theta_{ce} = 42.29°$

(b) $\alpha = 40°$ then $\alpha \geq \theta_{co}$, $\alpha \leq \theta_{ce}$

extraordinary ray escapes

(c) $\theta_{co} = 40.37°$ $\theta_{ce} = 40.08°$

(d) If we select $\alpha = 40.22°$

$\theta_{ce} \leq 40.22 \leq \theta_{co}$

extraordinary ray stays in plate.

4. *Double refraction: I.* For the case where the optic axis is perpendicular to the plane of incidence and in the plane of the interface, we have n_o and n_e independent of θ. The angle of refraction of the extraordinary ray can be calculated from the relation

$$\sin\theta_1 = n_o \sin\theta_2'' = n_e \sin\tilde{\theta}_2''$$

Calculate θ_2'' and $\tilde{\theta}_2''$ for $\theta_1 = 20°, 40°$, and $60°$ for both calcite ($n_o = 1.658$; $n_e = 1.486$) and quartz ($n_o = 1.544$; $n_e = 1.553$). Compare the values for calcite and quartz.

Solution

Calcite

θ_1	$\theta_2''\ (n_o)$	$\tilde{\theta}_2''\ (n_e)$
20	11.90°	13.31
40	22.81	25.63
60	31.49	35.65

Quartz

θ_1	$\theta_2''\ (n_o)$	$\tilde{\theta}_2''\ (n_e)$
20	12.80	12.72
40	24.60	24.45
60	34.12	33.89

5. *Double refraction: II.* For the case where the optic axis is in the plane of the interface and in the plane of incidence, determine graphically the angle of refraction $\tilde{\theta}_2''$ for calcite and quartz for $\theta_1 = 20°, 40°$, and $60°$. See problem 4 for the values of n_o and n_e, and use similar formulas.

5.5 Solution

calcite $n_o = 1.658$

$n_e = 1.486$

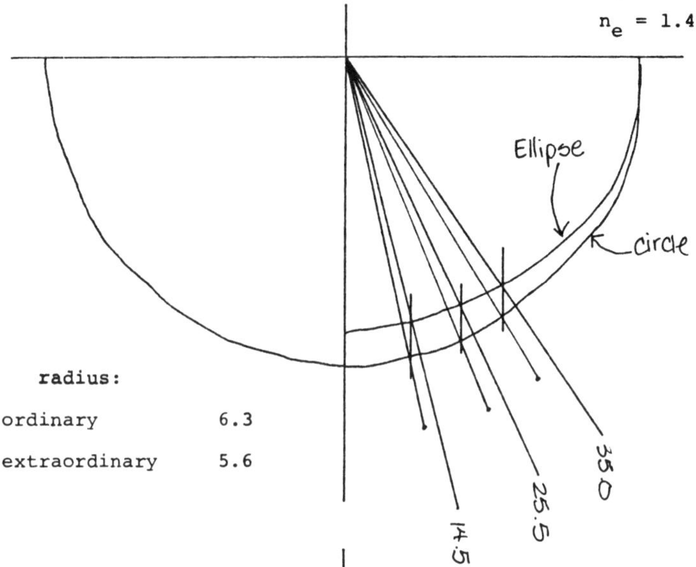

radius:

for ordinary 6.3

for extraordinary 5.6

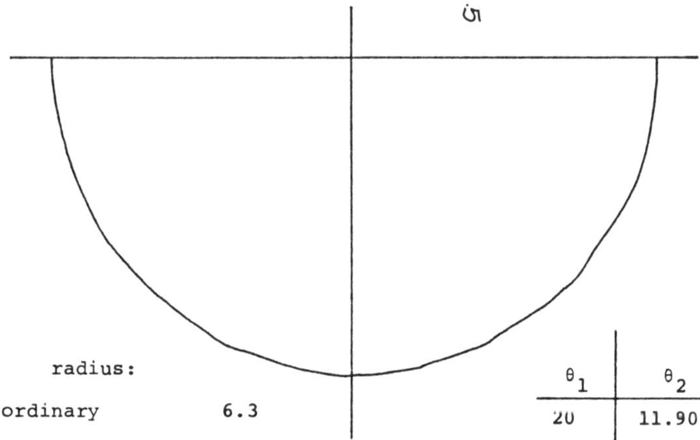

radius:

for ordinary 6.3

for extraordinary 6.3

θ_1	θ_2	Estimated $\tilde{\theta}_2''$
20	11.90	14.5
40	22.81	25.5
60	31.49	35.0

quartz $n_o = 1.544$

$n_e = 1.553$

does not work

with this scale

6. *Wollaston prism.* Consider a Wollaston prism with an angle of 20° (see accompanying figure). Calculate the angles α_1 and α_2 between the normal of the prism faces and the emerging ordinary and extraordinary waves. The prism is made of calcite ($n_o = 1.658$ and $n_e = 1.486$).

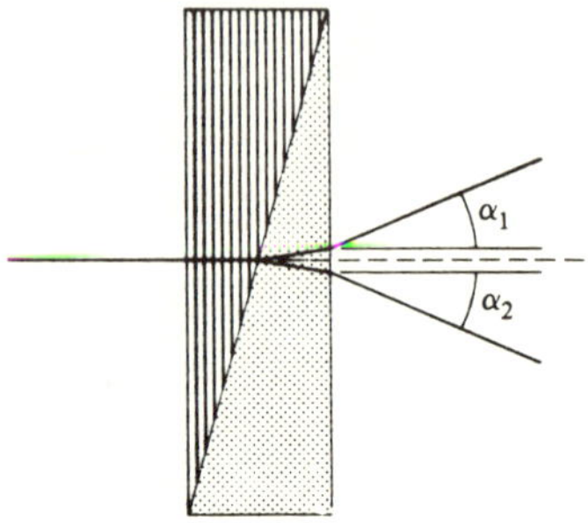

Solution

o-component in first crystal is e component in second and vice versa

For the e-ray in the first crystal

First interface $n_e \sin\theta = n_o \sin\theta_o$

$1.486 \sin 20 = 1.658 \sin\theta_o$ $\theta_o = 17.85$

$\alpha_2' = -2.15°$

Second interface $n_o \sin\alpha_2' = \sin\alpha_2$ $\alpha_2 = -3.57°$

For the o ray in the first crystal

First interface $n_o \sin\theta = n_e \sin\theta_e$

$1.658 \sin 20 = 1.486 \sin\theta_e$ $\theta_e = 22.43$

$\alpha_1' = +2.43$

Second interface $1.486 \sin 2.43 = \sin\alpha_1$ $\alpha_1 = 3.61$

7. *Kerr effect.* For carbon disulfide, the Kerr constant $K = 3.5 \times 10^{-12}$ cm/V^2; for chloroform, $K = -3.8 \times 10^{-12}$ cm/V^2. As a result, one of these compounds has $n_\| > n_\perp$. The velocity of the waves in the medium is $v = c/n$.
   a. For which compound do we have a higher velocity for the wave vibrating parallel to the E field compared to the wave vibrating perpendicular?
   b. Calculate $\Delta n_K = n_\| - n_\perp$ for nitrobenzene. Assume that $V = 3 \times 10^4$ V, $d =$ 1 cm, $\lambda = 0.5 \times 10^{-4}$ cm, and $K = 224 \times 10^{-12}$ cm/V^2.

Solution

(a) $n_\| - n_\perp = K \lambda_o \frac{V^2}{d^2}$

if K is neg. $n_\perp > n_\|$

if K is pos. $n_\perp < n_\|$

since $v = \frac{c}{n}$, if $n_{\parallel} < n_{\perp}$ then $v_{\parallel} > v_{\perp}$

answer: chloroform

(b) $\Delta n_K = n_{\parallel} - n_{\perp}$ for nitrobenzene

$= 1.008 \times 10^{-5}$

8. *Linear Pockels effect.* The difference between the refractive indices for vibrations along the X' an Y' axes is $\Delta n_p = -n_o^3 r_{63} E_Z$.
   a. Estimate n_o and r_{63} for CsDA and KDP from Figure 5.25. for what λ?
   b. For $\lambda_0 = 0.5 \times 10^{-4}$ cm and $E_Z = 3 \times 10^3$ V/cm, calculate Δn_p for CsDA and KDP.

Solution

(a) CsDA $n_o \simeq 1.56$

KDP $n_o \simeq 1.51$

CsDA $r_{63} \simeq 20 \times 10^{-10} \frac{cm}{V}$

KDP $r_{63} \simeq 11 \times 10^{-10} \frac{cm}{V}$

(b) CsDA $\Delta n_p = -2.28 \times 10^{-5}$

KDP $\Delta n_p = -1.14 \times 10^{-5}$

9. *Wave equation.* Show that we obtain the wave equation 5.27,

$$\nabla(\nabla \cdot \mathbf{E}) - \nabla^2 \mathbf{E} = -\mu_0 \frac{\partial^2 \mathbf{D}}{\partial t^2}$$

by taking the curl of equation 5.21,

$$\nabla \times \mathbf{E} = -\frac{\partial \mathbf{B}}{\partial t}$$

and using the relation

$$\nabla \times (\nabla \times \mathbf{A}) = \nabla(\nabla \cdot \mathbf{A}) - \nabla^2 \mathbf{A}$$

Solution: We have for curl $\vec{v} = \nabla \times \vec{v}$

$$= \left(\frac{\partial v_Z}{\partial Y} - \frac{\partial v_Y}{\partial Z} \right) \vec{i} + \left(\frac{\partial v_X}{\partial Z} - \frac{\partial v_Z}{\partial X} \right) \vec{j} + \left(\frac{\partial v_Y}{\partial X} - \frac{\partial v_X}{\partial Y} \right) \vec{k}$$

Therefore we have

$$\text{curl } (\nabla \times \vec{E}) = \nabla \times \nabla \times \vec{E} = \nabla \times \left(-\frac{\partial \vec{B}}{\partial t}\right)$$

The left side is

$$\nabla(\nabla \cdot \vec{E}) - \nabla^2 \vec{E}$$

The right side is $-\frac{\partial}{\partial t}(\nabla \times \vec{B})$, or with eq. 5.22

$$-\frac{\partial}{\partial t} \mu_o \frac{\partial \vec{D}}{\partial t} = -\mu_o \frac{\partial^2 \vec{D}}{\partial t^2}$$

therefore we have $\nabla(\nabla \cdot \vec{E}) - \nabla^2 \vec{E} = -\mu_o \frac{\partial^2 \vec{D}}{\partial t^2}$

Note ∇^2 is called the Laplacian in X,Y,Z coordianates

we have $\nabla^2 = \frac{\partial^2}{\partial X^2} + \frac{\partial^2}{\partial Y^2} + \frac{\partial^2}{\partial Z^2}$

10. *Equivalence of $\partial/\partial X$ and multiplication by ik_X.* Show that for plane waves we have the equivalence of the operations $\nabla \times \mathbf{E}$ with multiplication by $i(\mathbf{k} \cdot \mathbf{E})$ (equation 5.30). Write the expressions out in differential form and then do the substitution.

Solution:

$$\nabla \times \vec{E} = \begin{vmatrix} \vec{i} & \vec{j} & \vec{k} \\ \frac{\partial}{\partial X} & \frac{\partial}{\partial Y} & \frac{\partial}{\partial Z} \\ E_X & E_Y & E_Z \end{vmatrix} = \left(\frac{\partial E_Z}{\partial Y} - \frac{\partial E_Y}{\partial Z}\right)\vec{i} + \left(\frac{\partial E_X}{\partial Z} - \frac{\partial E_Z}{\partial X}\right)\vec{j} + \left(\frac{\partial E_Y}{\partial X} - \frac{\partial E_X}{\partial Y}\right)\vec{k}$$

Substitution:

$$(ik_Y E_Z - ik_Z E_Y)\vec{i} + (ik_Z E_X - ik_X E_Z)\vec{j}$$

$$+ (ik_X E_Y - ik_Y E_X)\vec{k} = i\,(\vec{k} \times \vec{E})$$

11. *Equation for* **E** *and* **k** *in the crystal.*
    a. Use

$$-\mathbf{k}(\mathbf{k} \cdot \mathbf{E}) + \mathbf{E}(\mathbf{k} \cdot \mathbf{k}) = -\mu_0 \frac{\partial^2 \mathbf{D}}{\partial t^2}$$

and

$$\mu_0 \frac{\partial^2 D_X}{\partial t^2} = \mu_0 \varepsilon_1 \frac{\partial^2 E_X}{\partial t^2}$$

$$\mu_0 \frac{\partial^2 D_Y}{\partial t^2} = \mu_0 \varepsilon_1 \frac{\partial^2 E_Y}{\partial t^2}$$

$$\mu_0 \frac{\partial^2 D_Z}{\partial t^2} = \mu_0 \varepsilon_2 \frac{\partial^2 E_Z}{\partial t^2}$$

and

$$\mu_0 \varepsilon_i = \frac{n_i^2}{c^2} \qquad (i = 1, 2)$$

and assume that $k_Y = 0$. Arrive at

$$-k_X(k_Z E_Z) + E_X k_Z^2 - n_1^2 \left(\frac{\omega}{c}\right)^2 E_X = 0$$

$$E_Y(k_X^2 + k_Z^2) - n_1^2 \left(\frac{\omega}{c}\right)^2 E_Y = 0$$

$$-k_Z k_X E_X + E_Z k_X^2 - n_2^2 \left(\frac{\omega}{c}\right)^2 E_Z = 0$$

b. Write the equations obtained in (a) as

$$a_{11} E_X + a_{12} E_Z + 0 E_Y = 0$$
$$a_{21} E_X + a_{22} E_Z + 0 E_Y = 0$$
$$0 E_X + 0 E_Z + a_{33} E_Y = 0$$

The determinant of this system of equations must vanish. Show that we get equation 5.38 for the determinant.

Solution (a)

$$-k_X(k_X E_X + k_Y E_Y + k_Z E_Z) + E_X(k_X^2 + k_Y^2 + k_Z^2) \; -\mu_o \varepsilon_1 \frac{\partial^2 E_X}{\partial t^2}$$

$$-k_Y(k_X E_X + k_Y E_Y + k_Z E_Z) + E_Y(k_X^2 + k_Y^2 + k_Z^2) = -\mu_o \varepsilon_1 \frac{\partial^2 E_Y}{\partial t^2}$$

$$-k_Z(k_X E_X + k_Y E_Y + k_Z E_Z) + E_Z \; (k_X^2 + k_Y^2 + k_Z^2) = -\mu_o \varepsilon_2 \frac{\partial^2 E_Z}{\partial t^2}$$

all terms containing k_Y are zero

$\frac{\partial^2 E_X}{\partial t^2} = -\omega^2 E_X$, same for E_Y and E_Z

therefore we have

$$-k_X(k_X E_X + k_Z E_Z) + E_X(k_X^2 + k_Z^2) = +\mu_o \varepsilon_1 \omega^2 E_X$$

$$E_Y(k_X^2 + k_Z^2) = +\mu_o \varepsilon_1 \omega^2 E_Y$$

$$-k_Z(k_X E_X + k_Z E_Z) + E_Z(k_X^2 + k_Z^2) = +\mu_o \varepsilon_2 \omega^2 E_Z$$

$$-k_X(k_Z E_Z) + E_X k_Z^2 - n_1^2 \frac{\omega^2}{c^2} E_X = 0$$

$$E_Y(k_X^2+k_Z^2) - n_1^2 \frac{\omega^2}{c^2} E_Y = 0$$

$$-k_Z(k_X E_X) + E_Z k_X^2 - n_2^2 \frac{\omega^2}{c^2} E_Z = 0$$

Solution (b)

$$(k_Z^2 - n_1^2 \frac{\omega^2}{c^2}) E_X - k_X k_Z E_Z + 0\, E_Y = 0$$

$$-k_Z k_X E_X + (k_X^2 - n_2^2 \frac{\omega^2}{c^2}) E_Z + 0\, E_Y = 0$$

$$0\, E_X + 0\, E_Z + [k_X^2 + k_Z^2 - n_1^2 \frac{\omega^2}{c^2}] E_Y = 0$$

The determinant is eq. 5.38.

---

6 Polarized Light

1. *Half-wave plate, quarter-wave plate.* The phase difference of an ordinary wave and an extraordinary wave after traveling the distance X is

$$\phi_X = |k_2 - k_1| X$$

where

$$k_2 = \frac{2\pi}{\lambda_0} n_e \quad \text{and} \quad k_1 = \frac{2\pi}{\lambda_0} n_o$$

The absolute value bars are introduced to make Δk or Δn positive. We can write ϕ_X as

$$\phi_X = \frac{2\pi}{\lambda_0} |n_e - n_o| X$$

where X is the coordinate of propagation. We assume that $\lambda = 0.5 \cdot 10^{-3}$ mm and that $n_o = 1.544$ and $n_e = 1.553$ for quartz and $n_o = 1.658$ and $n_e = 1.486$ for calcite.

a. Calculate the minimum thickness for a half-wave plate of quartz.
b. Calculate the minimum thickness for a half-wave plate of calcite.
c. Give the minimum thickness for a quarter-wave plate of quartz and for a quarter-wave plate of calcite.
d. Suppose that a 1 mm plate of quartz and 1.05 mm plate of calcite are available. How much material must be removed to make half-wave plates of quartz and calcite?
e. Suppose that the same plates are available as in (d). How much must be taken off to make quarter-wave plates of quartz and calcite?

Solution

(a) Half wave plate of quartz

$\pi = \frac{2\pi}{\lambda} \Delta n L \quad \rightarrow \quad L = .0278\text{mm}$

(b) Half wave plate of calcite L = .00145mm

(c) Quarter wave plate. Quartz = .0139mm

Calcite = .00073mm

(d) We need an odd number of plate thicknesses.

Half wave plate of quartz. L' = 1mm

$\frac{1}{.0278} = 35.97$

.97 of .0278 mm has to be taken off: .0269mm

Half wave plate of calcite:

$\frac{1.05}{.00145} = 724.14$

1.14 (.00145 mm) has to be taken off: 1.63×10^{-3}mm

(e) Quarter wave plate for quartz

$\frac{1}{.0139} = 71.94$

.94. .0139 mm has to be taken off: .0131mm

Quarter wave plate for calcite

$\frac{1.05}{.00073} = 1438.356$

1.356. (.00073mm) has to be taken off: 9.9×10^{-4}mm

2. *General form of the ellipse for elliptically polarized light*

a. Start with

$$E_Y = A \cos \alpha$$

and

$$E_Z = A(\cos \alpha \cos \phi - \sin \alpha \sin \phi)$$

These are equations 6.28 and 6.29, where α is time dependent. Eliminate the time dependence and give the general form of the ellipse. For $\cos \phi = \sin \phi = 1/\sqrt{2}$ and $A = 1$, we must get equation 6.31.

b. Use the transformation

$$E_Y = (\cos \phi)(E'_Y - E'_Z)$$
$$E_Z = (\cos \phi)(E'_Y + E'_Z)$$

and get the equation of an ellipse. For $\phi = \pi/4$, where $\cos \phi = \sin \phi = 1/\sqrt{2}$, we must get equation 6.33.

c. Write the ellipse for the special cases where $\phi = \pi/4$, $3\pi/4$, $5\pi/4$, and $7\pi/4$. Which pairs are the same? Compare with Table 6.2.

Solution

(a)

$$E_Y = A \cos \alpha$$

$$E_Z = A[\cos\alpha \cos\phi - \sin\alpha \sin\phi]$$

$$E_Z = E_Y \cos\phi - A\sqrt{1-\cos^2\alpha}\, \sin\phi$$

$$= E_Y \cos\phi - \sqrt{A^2 - A^2 \frac{E_Y^2}{A^2}}\, \sin\phi$$

$$E_Z = E_Y \cos\phi - \sqrt{A^2 - E_Y^2}\, \sin\phi$$

$$(E_Z - E_Y \cos\phi)^2 = (A^2 - E_Y^2) \sin^2\phi$$

$$E_Z^2 - 2E_ZE_Y \cos\phi + E_Y^2 \cos^2\phi = A^2 \sin^2\phi - E_Y^2 \sin^2\phi$$

$$E_Z^2 - 2E_ZE_Y \cos\phi + E_Y^2 (\cos^2\phi + \sin^2\phi) = A^2 \sin^2\phi$$

$$E_Z^2 - 2E_ZE_Y \cos\phi + E_Y^2 = A^2 \sin^2\phi$$

Check: $A = 1$, $\cos\phi = \sin\phi = \frac{1}{\sqrt{2}}$

$$E_Z^2 - \sqrt{2}\, E_ZE_Y + E_Y^2 = \frac{1}{2} \quad \text{that is 6.29}$$

(b)

Transformation

$$E_Y = \cos\phi\, (E_Y' - E_Z')$$

$$E_Z = \cos\phi\, (E_Y' + E_Z')$$

$$E_Y^2 = \cos^2\phi\, [E_Y'^2 - 2E_Y'E_Z' + E_Z'^2]$$

$$E_Z^2 = \cos^2\phi\, [E_Y'^2 + 2E_Y'E_Z' + E_Z'^2]$$

$$E_Z^2 + E_Y^2 = 2 \cos^2\phi\, [E_Y'^2 + E_Z'^2]$$

$$-2 \cos\phi\, E_ZE_Y = -2 \cos^3\phi\, (E_Y'^2 - E_Z'^2)$$

$$2 \cos^2\phi\, [E_Y'^2 + E_Z'^2] - 2 \cos^3\phi\, [E_Y'^2 - E_Z'^2] = A^2 \sin^2\phi$$

$$E_Y'^2 (2 \cos^2\phi - 2 \cos^3\phi) + E_Z'^2 (2 \cos^2\phi + 2 \cos^3\phi) = A^2 \sin^2\phi$$

$$\frac{E_Y'^2}{2 \cos^2\phi + 2 \cos^3\phi} + \frac{E_Z'^2}{2 \cos^2\phi - 2 \cos^3\phi} = \frac{A^2 \sin^2\phi}{(2 \cos^2\phi)^2 - (2 \cos^3\phi)^2}$$

check for $\phi = \frac{\pi}{4}$, $\cos\phi = \sin\phi = \frac{1}{\sqrt{2}}$ and $A = 1$

$$\frac{E_Y'^2}{1 + \frac{1}{\sqrt{2}}} + \frac{E_Z'^2}{1 - \frac{1}{\sqrt{2}}} = 1$$

(c)

$$\phi = \frac{\pi}{4} \quad \cos\phi = \sin\phi = \frac{1}{\sqrt{2}}$$

$$\frac{E_Y'^2}{1 + \frac{1}{\sqrt{2}}} + \frac{E_Z'^2}{1 - \frac{1}{\sqrt{2}}} = 1$$

$$\phi = \frac{3\pi}{4} \quad \cos\phi = \frac{-1}{\sqrt{2}}, \quad \sin\phi = \frac{1}{\sqrt{2}}$$

$$\frac{E_Y'^2}{1 - \frac{1}{\sqrt{2}}} + \frac{E_Z'^2}{1 + \frac{1}{\sqrt{2}}} = 1$$

$$\phi = \frac{5\pi}{4} \quad \cos\phi = \frac{-1}{\sqrt{2}}, \quad \sin\phi = \frac{-1}{\sqrt{2}}$$

$$\frac{E_Y'^2}{1 - \frac{1}{\sqrt{2}}} + \frac{E_Z'^2}{1 + \frac{1}{\sqrt{2}}} = 1$$

$$\phi = \frac{7\pi}{4} \quad \cos\phi = \frac{1}{\sqrt{2}}, \quad \sin\phi = -\frac{1}{\sqrt{2}}$$

$$\frac{E_Y'^2}{1 + \frac{1}{\sqrt{2}}} + \frac{E_Z'^2}{1 - \frac{1}{\sqrt{2}}} = 1$$

3. *Phase difference between internally reflected polarized components.* We want to calculate the phase difference δ between the internally reflected components of total internal reflection. We are interested in the region $\theta_c \leq \theta \leq 90°$. We write Fresnel's formulas for the case of internal total reflection:

$$r_{E_\perp} = \frac{\cos\theta - i\sqrt{\sin^2\theta - (n_2/n_1)^2}}{\cos\theta + i\sqrt{\sin^2\theta - (n_2/n_1)^2}}$$

$$r_{E_\parallel} = \frac{-(n_2/n_1)^2\cos\theta + i\sqrt{\sin^2\theta - (n_2/n_1)^2}}{(n_2/n_1)^2\cos\theta + i\sqrt{\sin^2\theta - (n_2/n_1)^2}}$$

In the region of total internal reflection, we know that $|r_{E_\perp}| = 1$ and $|r_{E_\parallel}| = 1$. The phase difference Δ between the two components can be expressed as

$$\frac{r_{E_\parallel}}{r_{E_\perp}} = \frac{|r_{E_\parallel}|}{|r_{E_\perp}|}e^{i\Delta} = \frac{(-(n_2/n_1)^2\cos\theta + i\sqrt{\qquad})/((n_2/n_1)^2\cos\theta + i\sqrt{\qquad})}{\cos\theta - i\sqrt{\qquad}/\cos\theta + i\sqrt{\qquad}}$$

a. Show that we have

$$e^{i\Delta} = \frac{-\sin^2\theta + i\cos\theta\sqrt{\sin^2\theta - (n_2/n_1)^2}}{\sin^2\theta + i\cos\theta\sqrt{\sin^2 - (n_2/n_1)^2}}$$

To do this, we multiply by

$$\sin^2\theta - i\cos\theta\sqrt{\qquad}$$

to be able to get the real and imaginary parts:

$$e^{i\Delta} = \frac{-(-\sin^2\theta + i\cos\theta\sqrt{\qquad})^2}{(\sin^2\theta + i\cos\theta\sqrt{\qquad})(\sin^2\theta - i\cos\theta\sqrt{\qquad})}$$

b. Now call the denominator A, because it is a real number, and show that we have

$$A^2 e^{+i\pi} e^{i\Delta} = (\sin^2\theta + i\cos\theta\sqrt{\qquad})^2$$

c. Show that we get

$$\tan\frac{\Delta}{2} = \frac{\sin^2\theta}{\cos\theta\sqrt{\sin^2\theta - (n_2/n_1)^2}}$$

Solution

(a)

$$e^{i\delta} = \frac{-(\frac{n_2}{n_1})^2 \cos\theta + i\sqrt{\ }}{(\frac{n_2}{n_1})^2 \cos\theta + i\sqrt{\ }} \cdot \frac{\cos\theta + i\sqrt{\ }}{\cos\theta - i\sqrt{\ }}$$

$$= \frac{-(\frac{n_2}{n_1})^2 \cos^2\theta + i(\cos\theta)\sqrt{\ } - i(\frac{n_2}{n_1})^2(\cos\theta)\sqrt{\ } - (\sin^2\theta - (\frac{n_2}{n_1})^2)}{(\frac{n_2}{n_1})^2 \cos^2\theta + i(\cos\theta)\sqrt{\ } - i(\frac{n_2}{n_1})^2(\cos\theta)\sqrt{\ } + (\sin^2\theta - (\frac{n_2}{n_1})^2)}$$

It is $(\frac{n_2}{n_1})^2 \cos^2\theta + \sin^2\theta - (\frac{n_2}{n_1})^2 \cos^2\theta - (\frac{n_2}{n_1})^2 \sin^2\theta = (\sin^2\theta)(1-(\frac{n_2}{n_1})^2)$

and

$$e^{i\Delta} = \frac{-(\sin^2\theta)(1-(\frac{n_2}{n_1})^2) + i(\cos\theta)\sqrt{\ }(1-(\frac{n_2}{n_1})^2)}{(\sin^2\theta)(1-(\frac{n_2}{n_1})^2) + i(\cos\theta)\sqrt{\ }(1-(\frac{n_2}{n_1})^2)}$$

$$= \frac{-\sin^2\theta + i(\cos\theta)\sqrt{\ }}{\sin^2\theta + i(\cos\theta)\sqrt{\ }}$$

(b) We have $-1 = e^{-i\pi}$ and get

$$A^2 e^{-i\pi} e^{i\Delta} = (-\sin^2\theta + i(\cos\theta)\sqrt{\ })^2$$

(c)

$$e^{i\frac{(\Delta-\pi)}{2}} = -\frac{\sin^2\theta}{A} + i\frac{(\cos\theta)\sqrt{}}{A}$$

The tan of $\frac{\Delta-\pi}{2}$ is equal to the ratio of the imaginary part to the real part

$$\tan\frac{\Delta-\pi}{2} = \frac{(\cos\theta)\sqrt{\sin^2\theta-(\frac{n_2}{n_1})^2}}{\sin^2\theta}$$

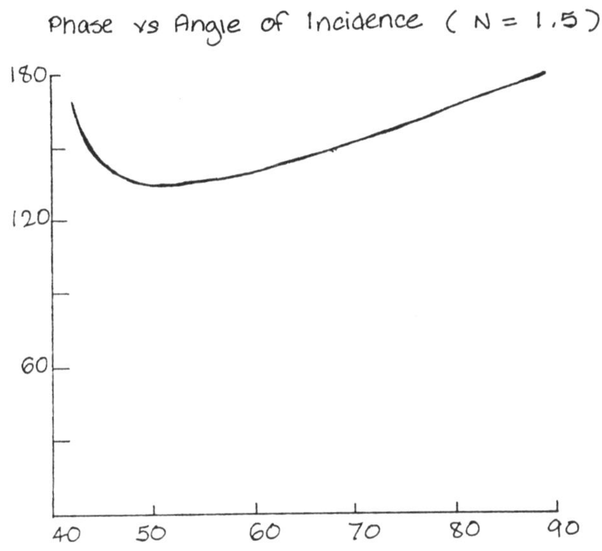

4. *Elliptically polarized light from total internal reflection: Jones vectors and matrices.* In Chapter 4, we calculated the reflection coefficient for total internal reflection. It turned out to be a complex number. A complex quantity can be written as $Re^{i\phi}$; that is, it can be described by its absolute value and its phase angle. If light polarized parallel and perpendicular to the plane of incidence is totally internally reflected, both components undergo a phase shift. If the phase difference is 90°, for example, we obtain circularly polarized light. We want to investigate what happens to the phase difference if linearly polarized incident light is internally reflected in a plane parallel plate one or more times (see accompanying figure). We are only interested in the part that is internally reflected and will ignore the light lost to the outside. For the reflected light polarized parallel and perpendicular to the plane of incidence,

$$r_{E_\parallel} = -\frac{db - ia}{db + ia} \quad \text{and} \quad r_{E_\perp} = \frac{b - ida}{b + ida}$$

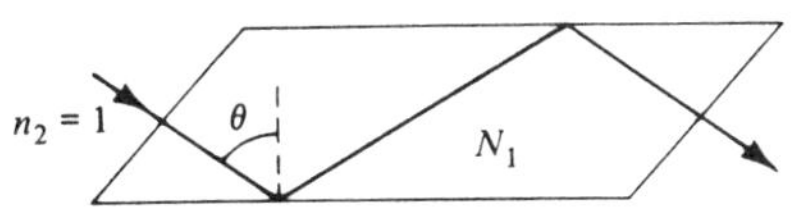

where

$$a = \sqrt{\left(\frac{N_1}{n_2}\right)^2 \sin^2\theta - 1}, \quad b = \cos\theta, \quad \text{and} \quad d = \frac{n_2}{N_1}$$

and where θ is the angle of incidence, $N_1 = N = 1.5$ (the index of refraction of the glass plate), and $n_2 = 1$ (the index of refraction outside). To find the relative phase difference after reflection, we rewrite one of the two expressions as

$$r_{E_\perp} = \frac{z_\perp^*}{z_\perp}$$

where

$$z_\perp = b + ida = Re^{i\beta}$$
$$z_\perp^* = b - ida = Re^{-i\beta}$$
$$\tan\beta = \frac{da}{b}$$

and we find $r_{E_\perp} = e^{-2i\beta}$.

a. Show that $r_{E_\parallel} = e^{-i(2\alpha-\pi)}$, where $\tan\alpha = a/bd$. For the relative phase difference after one reflection, we have

$$\phi = \arg r_{E_\parallel} - \arg r_{E_\perp} = 2\beta - 2\alpha + \pi$$

If many reflections are involved, or if different devices are involved, each producing a certain phase difference, we can develop a "bookkeeping" method to find out what the final phase difference is.

b. We now go back to our original problem. For one reflection, we obtained

$$\phi = 2\beta - 2\alpha + \pi \quad \text{or} \quad \beta - \alpha = \tfrac{1}{2}(\phi - \pi)$$

We are interested in finding a relation between the angle of incidence θ and the material parameters for the plate. To do this, we need tan $(\beta - \alpha)$ as

$$\tan(\beta - \alpha) = \frac{\tan\beta - \tan\alpha}{1 + \tan\beta\tan\alpha} = \tan\left(\frac{\phi - \pi}{2}\right) = q$$

Insert the expressions for $\tan\alpha$ and $\tan\beta$ in terms of a, b, and d, and obtain

$$q^2 = \frac{\cos^2\theta(N^2\sin^2\theta - 1)}{N^2\sin^4\theta}$$

c. Solve the resulting quadratic equation for $\sin^2\theta$ and get

$$\sin^2\theta = \frac{N^2+1}{2N^2(q^2+1)} \pm \sqrt{\left(\frac{N^2+1}{2N^2(q^2+1)}\right)^2 - \frac{1}{N^2(q^2+1)}}$$

d. We want to obtain, with one reflection, circularly polarized light. For the particular case of $\phi = 90^\circ$ and a glass with $N = 1.5$, find the angle of incidence θ from the equation in part (c); that is, show that $q^2 = 1$ and that the equation for $\sin^2\theta$ has no real solution.

e. Now we go to two reflections, that is, in Jones matrices,

$$\begin{pmatrix} 1 & 0 \\ 0 & e^{i\phi} \end{pmatrix}\begin{pmatrix} 1 & 0 \\ 0 & e^{i\phi} \end{pmatrix} = \begin{pmatrix} 1 & 0 \\ 0 & e^{2i\phi} \end{pmatrix} = \begin{pmatrix} 1 & 0 \\ 0 & e^{i\delta} \end{pmatrix}$$

where δ is the phase difference of 90°,

$$\delta = 2\phi = 4\beta - 4\alpha + 2\pi$$

or

$$\beta - \alpha = \frac{\delta}{4}$$

If we write

$$\tan(\beta - \alpha) = \tan\frac{\delta}{4} = q$$

we can use the formulas from before. Show that for glass ($N = 1.5$), the angle of incidence θ has values close to 50° and 53°. A device like this is called a Fresnel rhomb and produces circularly polarized light after two reflections.

Solution

(a) $r_{E_\parallel} = \dfrac{db-ia}{db+ia}$ $\qquad Z_\parallel = db + ia = Re^{i\alpha}$

$\qquad Z^*_\parallel = db - ia = Re^{-i\alpha}$

$$r_{E_\parallel} = \frac{Z^*_\parallel}{Z_\parallel} = -\frac{Re^{-i\alpha}}{Re^{i\alpha}} = e^{-i(2\alpha-\pi)}$$

$$\tan\alpha = \frac{a}{db}$$

(b) $\tan\alpha = \dfrac{a}{db}$, $\tan\beta = \dfrac{da}{b}$ $\quad b = \cos\theta \quad a = \sqrt{\left(\frac{N}{n}\right)^2 \sin^2\theta - 1}$

$d = n/N$

$$q = \frac{\frac{da}{b} - \frac{a}{db}}{1 + \left(\frac{a}{db}\right)\left(\frac{da}{b}\right)} = \frac{\frac{a}{b}\left(d - \frac{1}{d}\right)}{1 + (a/b)^2} = \frac{d - 1/d}{b/a + a/b}$$

$$q = \frac{ab\left(\frac{d^2-1}{d}\right)}{a^2+b^2} = \frac{\sqrt{N^2\sin^2\theta - 1}\ \cos\theta\ \left(\frac{1}{N} - N\right)}{N^2\sin^2\theta - 1 + \cos^2\theta}$$

$$-1+\cos^2\theta = -(1-\cos^2\theta) = -\sin^2\theta$$

$$q = \frac{\cos\theta\left(\frac{1}{N} - N\right)}{(N^2-1)\sin^2\theta}\sqrt{N^2\sin^2\theta - 1} = \frac{\cos\theta\,(1-N^2)\sqrt{N^2\sin^2\theta - 1}}{N(N^2-1)\sin^2\theta}$$

$$-q = \frac{\cos\theta\sqrt{N^2\sin^2\theta - 1}}{N\sin^2\theta}$$

$$q^2 = \frac{\cos^2\theta\,(N^2\sin^2\theta - 1)}{N^2\sin^4\theta}$$

(c) $\boxed{q^2N^2\sin^4\theta = \cos^2\theta\,(N^2\sin^2\theta - 1)} = N^2\sin^2\theta\cos^2\theta - \cos^2\theta$

$$q^2N^2\sin^4\theta = N^2\sin^2\theta(1-\sin^2\theta) - (1-\sin^2\theta)$$

$$= N^2\sin^2\theta - N^2\sin^4\theta - 1 + \sin^2\theta$$

$$\sin^4\theta(q^2N^2+N^2) - \sin^2\theta(N^2+1) + 1 = 0$$

$$N^2(q^2+1)\sin^4\theta - (N^2+1)\sin^2\theta + 1 = 0$$

$$\sin^2\theta = \frac{(N^2+1)}{2N^2(q^2+1)} \pm \frac{1}{2.N^2(q^2+1)}\sqrt{(N^2+1)^2-4N^2(q^2+1)}$$

$$\boxed{\sin^2\theta = \frac{N^2+1}{2N^2(q^2+1)} \pm \sqrt{\left[\frac{(N^2+1)}{2N^2(q^2+1)}\right]^2 - \frac{1}{N^2(q^2+1)}}}$$

(d)

$\phi = 90°$ $\quad N = 1.5 \quad$ $q = \tan\left(\frac{\phi-\pi}{2}\right) = \tan\left(\frac{90-180}{2}\right) = -1$

$q^2 = 1$

$$\sqrt{(N^2+1)^2 - 4N^2(q^2+1)} = \sqrt{(2.25+1)^2 - 4(2.25)(1+1)}$$

$$= \sqrt{10.56 - 18} \Rightarrow \text{imaginary}$$

$\sin^2\theta$ is complex => NO REAL θ

(e) $$\begin{pmatrix}1 & 0\\ 0 & e^{i\phi}\end{pmatrix}\begin{pmatrix}1 & 0\\ 0 & e^{i\phi}\end{pmatrix} = \begin{pmatrix}1 & 0\\ 0 & e^{2i\phi}\end{pmatrix} = \begin{pmatrix}1 & 0\\ 0 & 0^{i\delta}\end{pmatrix}$$

$\delta = 2\phi$

$\delta = 2\phi = 2[2\beta - 2\alpha + \pi] = 4\beta - 4\alpha + 2\pi = 4\beta - 4\alpha$

or $\quad \beta - \alpha = \delta/4$

then $\quad q = \tan(\beta-\alpha) = \tan\frac{\delta}{4} = \tan\left(\frac{90}{4}\right) = .414$

$q^2 = .172$

For $\quad N = 1.5$ and $q = .414 \qquad q^2 = .172$

$$\sin^2\theta = \frac{2.25+1}{2(2.25)(1.172)} \pm \frac{1}{2(2.25)(1.172)}\sqrt{(3.35)^2-4(2.25)(1.172)}$$

$$= .616 \pm .190\sqrt{.0145} = .616 \pm .023$$

$\theta_+ = 53.1° \qquad \theta_- = 50.4°$

5. *Half-voltage.* Calculate the half-voltage for ADP:

$$V_{1/2} = \frac{\lambda}{2n_o^3 r_{63}}$$

Assume that $\lambda = 0.5 \times 10^{-6}$ m, $n_o = 1.52$, and $r_{63} = 8.5 \times 10^{-12}$ m/V.

Solution: $$V_{1/2} = \frac{.5 \text{x} 10^{-6} \text{m}}{2 \text{x} (1.52)^3 \text{x} 8.5 \text{x} 10^{-12} \frac{\text{m}}{\text{V}}}$$

$$= \frac{.5}{2 \text{x} 3.51 \text{x} 8.5} \quad \frac{10^{-6+12}}{} \quad \text{V}$$

$$= 0.0084 \text{x} 10^6 \text{ V}$$

$$= 8400 \text{ Volt}$$

6. *Optical rotation in quartz.* Calculate the specific rotatory power α_s for quartz. Use

$$\lambda = 3968 \text{ Å}, n_R = 1.55810, \text{ and } n_L = 1.55821.$$

Solution: $$\alpha_s = \frac{\pi \text{x} (1.55810 - 1.55821)}{0.3968 \mu\text{m}} = \frac{\pi}{\lambda} [n_R - n_L]$$

$$|\alpha_s| = 8.7 \text{x} 10^{-4} \frac{1}{\mu\text{m}}$$

7. *Optical rotation of an active component in solution: sucrose.* The specific rotation $[\rho]$(degrees/(g/cm^3)) is defined for a specific wavelength λ as follows. Consider a 10 cm column of solution containing C g/cm^3 of active compound. $[\rho]$ is the angle of rotation of the plane of polarization for 10 cm traversal of the light if 1 g/cm^3 of active compound is contained. The angle α of rotation is therefore given as

$$\alpha = \frac{[\rho](\text{degrees}/(\text{g/cm}^3)) \cdot X_{[\text{cm}]} \cdot C_{[\text{g/cm}^3]}}{10 \text{ cm}}$$

where X is the distance of travel in centimeters, C is the concentration in g/cm^3 and $[\rho]$ is given for a specific wavelength.

Calculate for sucrose using the following information: $[\rho] = 66.5°$ [1/(g/cm^3)] (for $\lambda = 5893$ Å, sodium D line), $X = 20$ cm, and $C = 0.25$ g/cm^3.

Solution: $$\alpha = \frac{66.5 \frac{1°}{\frac{\text{gr}}{\text{cm}^3}} \text{x} 20 \text{cm} \text{x} 0.25 \frac{\text{g}}{\text{cm}^3}}{10 \text{ cm}}$$

$$= 33.2°$$

8. *Linear polarizers.* The following Jones matrices represent horizontal, vertical, and 45° polarizers, respectively:

$$\begin{pmatrix} 1 & 0 \\ 0 & 0 \end{pmatrix} \quad \begin{pmatrix} 0 & 0 \\ 0 & 1 \end{pmatrix} \quad \frac{1}{2}\begin{pmatrix} 1 & 1 \\ 1 & 1 \end{pmatrix}$$

Horizontal Vertical 45°

a. Apply these matrices to vector representation of horizontal, vertical, and 45° polarized light and give an interpretation of the result:

$$\begin{pmatrix} 1 \\ 0 \end{pmatrix} \quad \begin{pmatrix} 0 \\ 1 \end{pmatrix} \quad \begin{pmatrix} 1 \\ 1 \end{pmatrix}$$

Horizontal Vertical 45°

b. Show that the matrix

$$\frac{1}{2}\begin{pmatrix} 1 & -1 \\ -1 & 1 \end{pmatrix}$$

presents a polarizer of $-45°$. Operate on $+45°$ polarized light,

$$\begin{pmatrix} 1 \\ 1 \end{pmatrix}$$

and on $-45°$ polarized light,

$$\begin{pmatrix} -1 \\ +1 \end{pmatrix}$$

and give an interpretation.

c. Let light first be polarized horizontally, then pass through a $+45°$ polarizer, and then through a $-45°$ polarizer. What is the polarization of the emerging light?

Solution

(a)

$$\begin{pmatrix} 1 & 0 \\ 0 & 0 \end{pmatrix}\begin{pmatrix} 1 \\ 0 \end{pmatrix} = \begin{pmatrix} 1 \\ 0 \end{pmatrix} \qquad \text{horizontal}$$

$$\begin{pmatrix} 0 & 0 \\ 0 & 1 \end{pmatrix}\begin{pmatrix} 1 \\ 0 \end{pmatrix} = \begin{pmatrix} 0 \\ 0 \end{pmatrix} \qquad \text{no light}$$

$$\frac{1}{2}\begin{pmatrix} 1 & 1 \\ 1 & 1 \end{pmatrix}\begin{pmatrix} 1 \\ 0 \end{pmatrix} = \frac{1}{2}\begin{pmatrix} 1 \\ 1 \end{pmatrix} \qquad \text{45° polarized}$$

$$\begin{pmatrix} 1 & 0 \\ 0 & 0 \end{pmatrix}\begin{pmatrix} 0 \\ 1 \end{pmatrix} = \begin{pmatrix} 0 \\ 0 \end{pmatrix} \qquad \text{no light}$$

$$\begin{pmatrix} 0 & 0 \\ 0 & 1 \end{pmatrix}\begin{pmatrix} 0 \\ 1 \end{pmatrix} = \begin{pmatrix} 0 \\ 1 \end{pmatrix} \qquad \text{vertical}$$

$$\frac{1}{2}\begin{pmatrix} 1 & 1 \\ 1 & 1 \end{pmatrix}\begin{pmatrix} 0 \\ 1 \end{pmatrix} = \frac{1}{2}\begin{pmatrix} 1 \\ 1 \end{pmatrix} \qquad \text{45° polarized}$$

$$\begin{pmatrix} 1 & 0 \\ 0 & 0 \end{pmatrix} \begin{pmatrix} 1 \\ 1 \end{pmatrix} = \begin{pmatrix} 1 \\ 0 \end{pmatrix} \quad \text{horizontal}$$

$$\begin{pmatrix} 0 & 0 \\ 0 & 1 \end{pmatrix} \begin{pmatrix} 1 \\ 1 \end{pmatrix} = \begin{pmatrix} 0 \\ 1 \end{pmatrix} \quad \text{vertical}$$

$$\frac{1}{2}\begin{pmatrix} 1 & 1 \\ 1 & 1 \end{pmatrix} \begin{pmatrix} 1 \\ 1 \end{pmatrix} = \begin{pmatrix} 1 \\ 1 \end{pmatrix} \quad 45° \text{ polarized}$$

(b) $$\frac{1}{2}\begin{pmatrix} 1 & -1 \\ -1 & 1 \end{pmatrix} \begin{pmatrix} 1 \\ 1 \end{pmatrix} = \begin{pmatrix} 0 \\ 0 \end{pmatrix} \quad \text{no light}$$

$$\frac{1}{2}\begin{pmatrix} 1 & -1 \\ -1 & 1 \end{pmatrix} \begin{pmatrix} -1 \\ +1 \end{pmatrix} = \begin{pmatrix} -1 \\ +1 \end{pmatrix} \quad -45° \text{ polarized}$$

(c) $$\frac{1}{2}\begin{pmatrix} 1 & -1 \\ -1 & 1 \end{pmatrix} \frac{1}{2}\begin{pmatrix} 1 & 1 \\ 1 & 1 \end{pmatrix} \begin{pmatrix} 1 & 0 \\ 0 & 0 \end{pmatrix}$$

zero matrix

No light emerges, it does not matter what the state of polarization of the incident light is.

9. *Combination of polarizers: presented in matrices*
   a. What is the action of first a quarter-wave plate and then a half-wave plate?
   b. What is the action of first a half-wave plate and then a quarter-wave plate?
   c. What is the action of a half-wave plate, then a quarter-wave plate, and then another half-wave plate?

Solution

(a) $$\begin{pmatrix} 1 & 0 \\ 0 & -1 \end{pmatrix} \begin{pmatrix} 1 & 0 \\ 0 & i \end{pmatrix} = \begin{pmatrix} 1 & 0 \\ 0 & -i \end{pmatrix} \quad \text{left circular}$$

(b) $$\begin{pmatrix} 1 & 0 \\ 0 & i \end{pmatrix} \begin{pmatrix} 1 & 0 \\ 0 & -1 \end{pmatrix} = \begin{pmatrix} 1 & 0 \\ 0 & -i \end{pmatrix} \quad \text{left circular}$$

(c) $$\begin{pmatrix} 1 & 0 \\ 0 & -1 \end{pmatrix} \underbrace{\begin{pmatrix} 1 & 0 \\ 0 & i \end{pmatrix} \begin{pmatrix} 1 & 0 \\ 0 & -1 \end{pmatrix}}_{\begin{pmatrix} 1 & 0 \\ 0 & -i \end{pmatrix}} = \begin{pmatrix} 1 & 0 \\ 0 & +i \end{pmatrix} \quad \text{right circular}$$

7 Wave Theory in Dispersive Media and Optical Constants

1. *The forced and damped oscillators.*

A. *The forced oscillator.* Let us review the mechanical forced oscillator. We start with the free oscillator

$$m\ddot{x} + kx = 0$$

where m is the mass, k is the force constant, and x is the displacement, and $\ddot{x} = d^2x/dt^2$. We will call ω_0 the resonance frequency.

a. Use the trial solution $x = A\sin(\omega_0 t + \delta)$ and show how the frequency depends on the constant k and the mass m.

The free oscillator may be driven by a force $F_0 \sin \omega t$, and we must consider the equation

$$m\ddot{x} + m\omega_0^2 x = F_0 \sin \omega t.$$

b. Use the trial solution $x = A\sin \omega t$ and find the solution for the above equation.

At the resonance frequency $\omega = \omega_0$, the amplitude is infinite. Therefore, we will now consider the more realisitic damped oscillator.

B. *The damped oscillator.* To avoid the above difficulty, we introduce a damping term and consider the free damped oscillator

$$m\ddot{x} + b\dot{x} + kx = 0$$

c. Use the trial solution $x = Ae^{i\bar{\omega}_0 t}$ and give the dependence of $\bar{\omega}_0$ on the constants m, b, and k.

d. Write the solution in the form $x = Ae^{\alpha t + i\beta t}$ and discuss α and β. What is their significance?

We now turn to the damped forced oscillator equation

$$m\ddot{x} + b\dot{x} + kx = F_0 e^{i\omega t}$$

e. Use the trial solution $x = Ae^{i\omega t}$ and introduce $\omega_0^2 = k/m$ of the free oscillator. Show that we get

$$A = \frac{F_0}{m(\omega_0^2 - \omega^2) + i\omega b}$$

Note that for $b = 0$ and $\omega \to \omega_0$, we have the same problem we had for the free osillator.

f. Rewrite the result in the form $x = \bar{A}e^{i(\omega t + \delta)}$. To determine $\bar{A}$, take the square root of the absolute value of x.

The phase angle δ is obtained by writing x (without the factor $e^{i\omega t}$) in complex form, that is, multiply numerator and denominator by

$$m(\omega_0^2 - \omega^2) - i\omega b$$

and take the ratio of the imaginary part over the real part.

g. Show that

$$\tan\delta = \frac{-\omega b}{m(\omega_0^2 - \omega^2)}$$

and give the final solution Ans. see 2a

h. Sketch the amplitude and phase angle as function of ω.

Solution

(a)

$$\dot{x} = A \cos(\omega_o t + \delta)\cdot\omega_o$$

$$\ddot{x} = -A\omega_o^2 \sin(\omega_o t + \delta)$$

$$-mA\omega_o^2 + Ak = 0$$

$$\omega_o^2 = \frac{k}{m} \qquad \omega_o = \sqrt{\frac{k}{m}}$$

(b)

$$-mA\omega^2 \sin \omega t + m\omega_o^2 A \sin \omega t$$

$$= F_o \sin \omega t$$

$$A = \frac{F_o}{m(\omega_o^2 - \omega^2)}$$

$$x = \frac{F_o}{m(\omega_o^2 - \omega^2)} \sin \omega t$$

(c)

$$-m\,\bar{\omega}_o^2 + b\, i\, \bar{\omega}_o + k = 0$$

$$\bar{\omega}_o^2 - i\, b\, \frac{\bar{\omega}_o}{m} - \frac{k}{m} = 0$$

$$\bar{\omega}_o = i\, \frac{b}{2}\, \frac{1}{m} \pm \sqrt{-\left(\frac{b}{2m}\right)^2 + \frac{k}{m}}$$

(d)

$$x = A\, e^{i\left[i\frac{b}{2m} \pm\sqrt{-\left(\frac{b}{2m}\right)^2+\frac{k}{m}}\right]t} = A\, e^{\alpha t + i\beta}$$

$$\alpha = -\frac{b}{2m} \qquad \beta = \pm\sqrt{-\left(\frac{b}{2m}\right)^2+\frac{k}{m}}$$

$\alpha \rightarrow$ exponential decay

$\beta \rightarrow$ oscillation

(e)

$$-A\,\omega^2 m + i\,\omega\, b\, A + k\, A = F_o$$

$$A = \frac{F_o}{-\omega^2 m + k + i\,\omega\, b} = \frac{F_o}{m(\omega_o^2-\omega^2) + i\,\omega\, b}$$

(f)

$$\bar{A} = \frac{F_o}{\sqrt{m^2(\omega_o^2 - \omega^2)^2 + \omega^2 b^2}}$$

(g) $$\frac{F_o[m(\omega_o^2 - \omega^2) - i\,\omega\, b]}{[m(\omega_o^2 - \omega^2)]^2 + \omega^2 b^2} \rightarrow \frac{-\,\omega\, b}{m(\omega_o^2 - \omega^2)} = \tan\delta$$

We indicate at the triangle the functional values

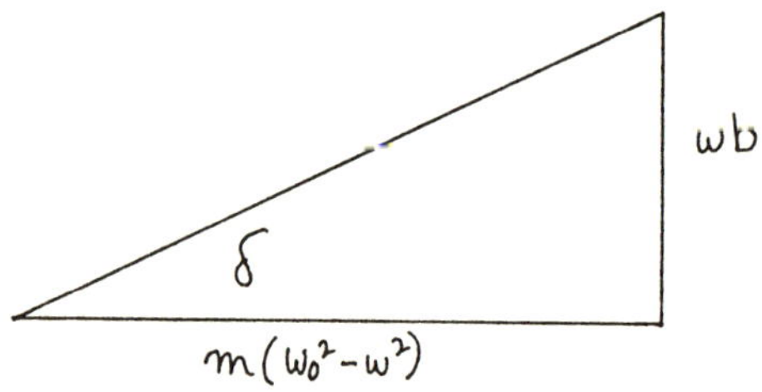

The final solution is now

$$x = \frac{F_o}{\sqrt{m^2(\omega_o^2 - \omega^2)^2 + \omega^2 b^2}}\, e^{i(\omega t + \delta)}$$

(h)

We sketch the amplitude as function of ω

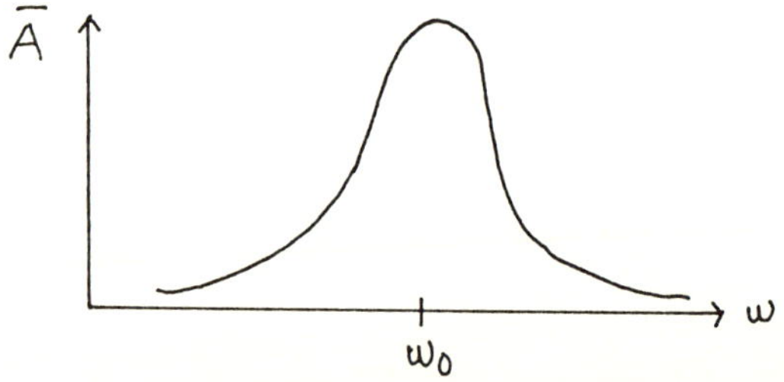

and sketch the phase angle δ

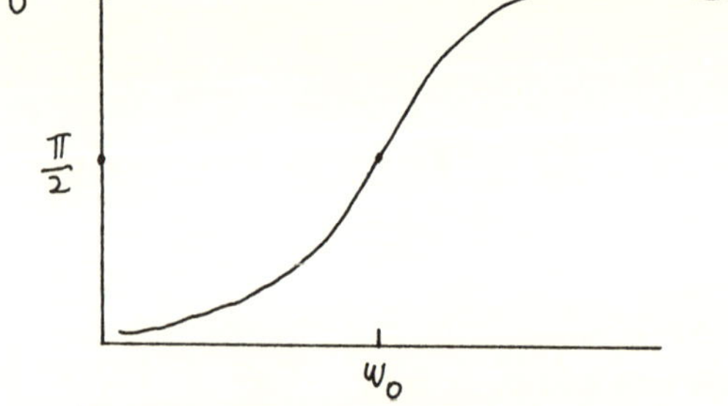

2. *Power considerations of the damped oscillator.* The average power may be obtained by multiplication of the (complex) force by $\dot{x}$ (where $dx/dt = \dot{x}$): and taking the time average

$$m\ddot{x} + b\dot{x} + kx = F$$
$$\overline{P_{av}} = \overline{(F\dot{x})} = m\overline{\dot{x}\ddot{x}} + b\overline{\dot{x}^2} + k\overline{x\dot{x}}$$

The averages of $\overline{\dot{x}\ddot{x}}$ and $\overline{x\dot{x}}$ are both zero.
Therefore, we have

$$\overline{P_{av}} = b\overline{\dot{x}^2}$$

a. Calculate $\overline{\dot{x}^2}$ by using the real part of the solution above, that is,

$$x = \frac{F_0}{\sqrt{m^2(\omega_0^2 - \omega^2)^2 + \omega^2 b^2}} \cos(\omega t + \delta)$$

and observe that the time average of $\sin^2(\omega t + \delta)$ is 1/2, see problem 1, Chapter 2.

We now consider the case of different dampings. If $b = 0$, no power is taken in by the oscillator.

b. If we plot $\overline{P_{av}}$ for large and small b, we obtain two curves like this:

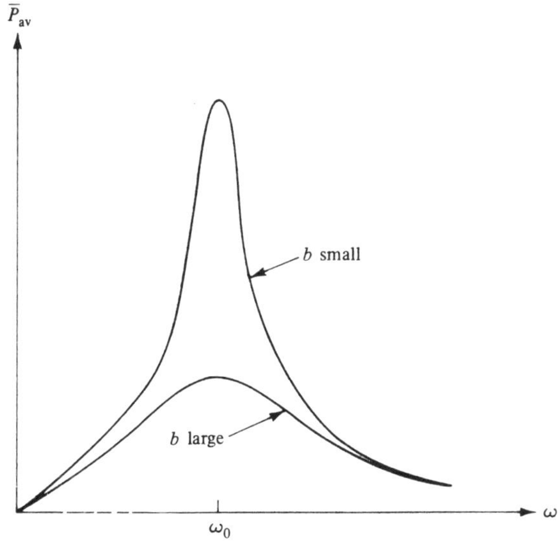

Solution

(a)
$$\dot{x} = -A \sin(\omega t + \delta)\omega$$

$$A = \frac{F_o}{(m^2(\omega_o^2 - \omega^2)^2 + \omega^2 b^2)^{\frac{1}{2}}}$$

$$\overline{\dot{x}^2} = \omega^2 A^2 \sin^2(\omega t + \delta) = \frac{\omega^2}{2} A^2$$

$$\overline{\dot{x}^2} = \frac{\omega^2}{2} \frac{F_o^2}{(\sqrt{m^2(\omega_o^2 - \omega^2)^2 + \omega^2 b^2})^2}$$

$$\overline{P_{av}} = \frac{\frac{\omega^2}{2} F_o^2}{m^2(\omega_o^2 - \omega^2)^2 + \omega^2 b^2}$$

3. *Rubens-Hagen approximation for metals.* For metals, we have the approximation

$$\bar{n} \simeq \bar{K} \simeq \sqrt{\frac{\sigma}{2\varepsilon_0 \omega}}$$

a. Show that the reflectivity at normal incidence

$$R = \frac{(\bar{n} - 1)^2 + \bar{K}^2}{(\bar{n} + 1)^2 + \bar{K}^2}$$

may be written as

$$R \simeq 1 - 2\frac{1}{\sqrt{\sigma/2\varepsilon_0 \omega}}$$

if we neglect terms in

$$\left(\frac{1}{\sqrt{\sigma/2\varepsilon_0 \omega}}\right) \text{ with respect to } \left(\frac{1}{\sqrt{\sigma/2\varepsilon_0 \omega}}\right)^2$$

This is called the *Rubens-Hagen approximation.*

b. Calculate $\sigma/2\varepsilon_0\omega$ for infrared frequencies (use $\lambda = 1$ mm and show how close R is to the value of 1 for copper).

$$\varepsilon_0 = 8.85 \times 10^{-12} \frac{C^2}{Nm} \quad \text{and} \quad \sigma = 6 \times 10^7 \,(Am)^{-1}$$

c. What do we obtain for $\omega \to 0$?

Solution

(a)

$$R = \frac{(1 - \frac{1}{\bar{n}})^2 + 1}{(1 + \frac{1}{\bar{n}})^2 + 1} \rightarrow \frac{1 - \frac{2}{\bar{n}} \;\ldots.\; + 1}{1 + \frac{2}{\bar{n}} \;\ldots.\; + 1}$$

$$\rightarrow \frac{1 - \frac{1}{\bar{n}}}{1 + \frac{1}{\bar{n}}} \;\ldots. \rightarrow (1 - \frac{1}{\bar{n}})(1 - \frac{1}{\bar{n}})$$

$$\rightarrow 1 - \frac{2}{\bar{n}} \;\ldots.$$

(b)

$$R = 1 - \frac{2}{\bar{n}} \qquad \varepsilon_o = 8.85 \times 10^{-12} \frac{C^2}{Nm^2}$$

$$\bar{n} = \sqrt{\frac{\sigma}{2\varepsilon_o \omega}} \qquad \sigma = 6 \times 10^7 \; (Am)^{-1}$$

$$\bar{n} = \sqrt{\frac{(6 \times 10^7)}{2 \cdot (8.85 \times 10^{-12})(1.88 \times 10^{12})} \frac{\text{Nm}^2 \cdot \text{s}}{\text{C}^2 \text{ A m}}}$$

$$\bar{n} = \sqrt{1.803 \times 10^6} = 1.343 \times 10^3 = 1343$$

$$R = 1 - \frac{2}{\bar{n}} = 1 - 0.0015 = 0.9985$$

(c)

$$R \to 1$$

4. *Ray deviation by a prism.* In Chapter 1, we discussed the angle of deviation of a prism. Now we want to obtain the angle for minimum deviation. For the prism, we have

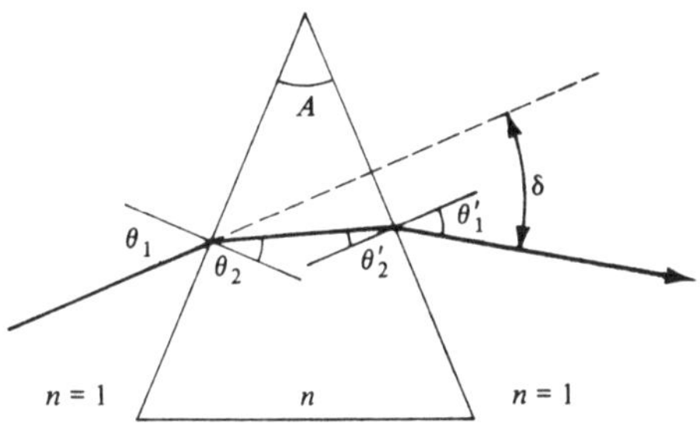

The refractive indices are indicated. We have

$$\sin\theta_1 = n\sin\theta_2$$
$$n\sin\theta_2' = \sin\theta_1'$$
$$\delta = \theta_1 + \theta_1' - A$$
$$A = \theta_2 + \theta_2'$$

These equations are invariant under the change from unprimed angles to primed angles. This suggests that we may have an optimum case for $\theta_1 = \theta_1'$, that is, a symmetric path. We now want to show that δ is minimum for the symmetric path. According to Sommerfeld we do this as follows:

a. Eliminate θ_1' and θ_2 from the four equations.

b. Differentiate with respect to θ_1 and θ_2' and show that we get

$$(\cos\theta_1)\,d\theta_1 + n\cos(A - \theta_2')\,d\theta_2' = 0$$

$$(\cos\theta_2')\,d\theta_2' + \frac{1}{n}\cos(\delta + A - \theta_1)\,d\theta_1 = 0$$

The two equation can be considered as a system of two homogeneous equations for $d\theta_1$ and $d\theta_2'$. To obtain a nontrivial solution, the determinant must vanish?

c. Show that the condition for the determinant to vanish is

$$\cos\theta_1\cos\theta_2' = \cos(A - \theta_2')\cos(\delta + A - \theta_1)$$

d. Show that one possible solution is

$$\theta_1 = \theta_1' = \frac{\delta + A}{2}, \quad \theta_2 = \theta_2' = \frac{A}{2}$$

and that the other possible solution is not useful to determine the significance of this symmetric path.

e. Calculate δ for $n = 1.5$, and $A = 30°$ and determine $\theta_1 = \theta_1'$.

f. Calculate δ for several values of θ_1 near the symmetric position $\theta_1 = 22.8°$ by filling in the following table:

θ_1	θ_2	θ_2'	θ_1'	δ
26°				
24°				
22°				
20°				
18°				

Note that δ goes through a miminum. The fact that δ has its minimum value for the symmetric path may be used in the determination of the refractive index.

g. Show that the refractive index at minimum derivation is

$$n = \frac{\sin\frac{1}{2}(\delta + A)}{\sin\frac{1}{2}A}$$

Solution

(a)

$$\sin\theta_1 = n\sin\theta_2 = n\sin(A-\theta_2')$$

$$n\sin\theta_2' = \sin\theta_1' = \sin(\delta-\theta_1+A)$$

(b)

$$\cos\theta_1(d\theta_1) = n\cos(A-\theta_2')(d\theta_2')$$

or

$$\cos\theta_1(d\theta_1) + n\cos(A-\theta_2')(d\theta_2') = 0$$

$$n\cos\theta_2'(d\theta_2') = -\cos(\delta-\theta_1+A)(d\theta_1)$$

or

$$\cos\theta_2'(d\theta_2') + \frac{1}{n}\cos(\delta-\theta_1+A)(d\theta_1) = 0$$

(c)

$$\begin{vmatrix} \cos\theta_1 & n\cos(A-\theta_2') \\ \frac{1}{n}\cos(\delta-\theta_1+A) & \cos\theta_2' \end{vmatrix} = 0$$

$$\cos\theta_1\cos\theta_2' = \cos(A-\theta_2')\cos(\delta-\theta_1+A)$$

(d)

$$\cos\frac{\delta+A}{2}\cos\frac{A}{2} = \cos\left(A - \frac{A}{2}\right)\cos\left(\delta+A - \frac{\delta+A}{2}\right)$$

For $\theta_1 = A - \theta_2'$, $\theta_2' = \delta + A - \theta_1$

$$\cos\theta_1 \cos\theta_2' = \cos\ [A-(\delta+A-\theta_1)]\ \cos\ [\delta+A-(A-\theta_2')]$$
$$= \cos(\theta_1-\delta)\ \cos(\theta_2' + \delta)$$

It follows that $\delta = 0$, that is: no deviation

(e) $n = 1.5$ $\quad A = 30°$ $\quad \theta_1 = \theta_1'$, $\theta_2 = \theta_2'$

$\sin\theta_1 = n\ \sin\theta_2 = n \sin\frac{A}{2} = 1.5 \sin 15°$

$\theta_1 = 22.8°$

$\delta = 2\theta - A = 2(22.8) - 30 = 15.7°$

(f)

θ_1	θ_2	θ_2'	θ_1'	δ	
26	17.0	13.0	19.7	15.7	
24	15.7	14.3	21.7	15.7	
22	14.5	15.5	23.6	15.6	← minimum
20	13.2	16.8	25.7	15.7	
18	11.9	18.1	27.8	15.8	

(g)

$$\sin\theta_1 = n \sin\theta_2$$

$$\sin\frac{\delta+A}{2} = n\ \sin\frac{A}{2}$$

$$\rightarrow \quad n = \frac{\sin\frac{\delta+A}{2}}{\sin\frac{A}{2}}$$

5. *Prism dispersion.* The prism is a useful dispersion device. If polychromatic light is incident upon a prism, the components of different wavelength will each be deviated, depending on the refractive index of each wavelength. In ultraviolet, visible, and infrared spectrometers, prisms are used to disperse radiation from a source and produce, with the use of slits, quasi-monochromatic bands of light. In this problem, we will be concerned with those properties of prism dispersion of interest to instrument designers.

A quantity of interest in the design of monochromators is the angular dispersion $d\delta/d\lambda$, which is the rate of change of deviation δ with wavelength. By writing

$$\frac{d\delta}{d\lambda} = \frac{d\delta}{dn}\left(\frac{dn}{d\lambda}\right) = G\frac{dn}{d\lambda}$$

we see that the angular dispersion is a product of a geometric factor ($G = d\delta/dn$) and the dispersion $dn/d\lambda$, a factor dependent only on the prism characteristics.

a. Consider an equilateral prism of side B and angle A. Show that for the special case of minimum deviation, the geometric factor becomes

$$G = \frac{d\delta}{dn} = \frac{B}{b}$$

where b is the beam diameter.

b. Show that the angular dispersion is

$$\frac{d\delta}{d\lambda} = \frac{B}{b}\left(\frac{dn}{d\lambda}\right)$$

Now consider the dispersion factor. In regions where no resonances occur, the index of refraction may be represented empirically by an expression of the form

$$n(\lambda) = c_1 + \frac{c_2}{\lambda^2} + \frac{c_3}{\lambda^4}$$

where c_1, c_2, and c_3 are constants.

c. Calculate the constants c_1, c_2, and c_3 for fused quartz. Take the values

$$n = 1.486 \quad \text{for} \quad \lambda = 3034 \text{ Å}$$
$$n = 1.463 \quad \text{for} \quad \lambda = 4800 \text{ Å}$$
$$n = 1.455 \quad \text{for} \quad \lambda = 7065 \text{ Å}$$

Solution

(a)

$$\delta = \theta_1 + \theta_1' - A$$

At minimum deviation $\theta_1 = \theta_1'$

$$\delta = 2\,\theta_1 - A$$

$$\frac{d\delta}{dn} = \frac{d}{dn}(2\,\theta_1 - A) = 2\,\frac{d\theta_1}{dn}$$

with $\quad \sin\theta_1 = n \sin\theta_2$

and $\quad \cos\theta_1\, d\,\theta_1 = dn \sin\theta_2$ (d θ_2 is about 0)

$$\frac{d\theta_1}{dn} = \frac{\sin\theta_2}{\cos\theta_1} \cdot \frac{a}{a}$$

$$\frac{d\delta}{dn} = 2\,\frac{d\theta_1}{dn} = \frac{2\sin\theta_2 \cdot a}{\cos\theta_1 \cdot a} = \frac{2\,a\sin(\frac{A}{2})}{b}$$

with $\quad a\cos\theta_1 = b$

and $\quad \cdot 2a\sin\frac{A}{2} = B$

$$\frac{d\delta}{dn} = \frac{B}{b}$$

(b)

$$\frac{d\delta}{dn} = \frac{B}{b} = \frac{d\delta}{d\lambda}\,\frac{d\lambda}{dn}$$

$$\rightarrow \frac{d\delta}{d\lambda} = \frac{B}{b}\,\frac{dn}{d\lambda}$$

(c)

$$1.486 = C_1 + \frac{C_2}{(3034)^2} + \frac{C_3}{(3034)^4}$$

$$1.463 = C_1 + \frac{C_2}{(4800)^2} + \frac{C_3}{(4800)^4}$$

$$1.455 = C_1 + \frac{C_2}{(7065)^2} + \frac{C_3}{(7065)^4}$$

$$C_1 + 10.86 \cdot 10^{-8}\ C_2 + 118 \cdot 10^{-16}\ C_3 = 1.486$$

$$C_1 + 4.34 \cdot 10^{-8}\ C_2 + 18.84 \cdot 10^{-16}\ C_3 = 1.463$$

$$C_1 + 2.00 \cdot 10^{-8}\ C_2 + 4.01 \cdot 10^{-16}\ C_3 = 1.455$$

$$C_1 = 1.448$$

$$C_2 = 3.3 \times 10^5\ \text{Å}^2$$

$$C_3 = 1.23 \times 10^{11}\ \text{Å}^4$$

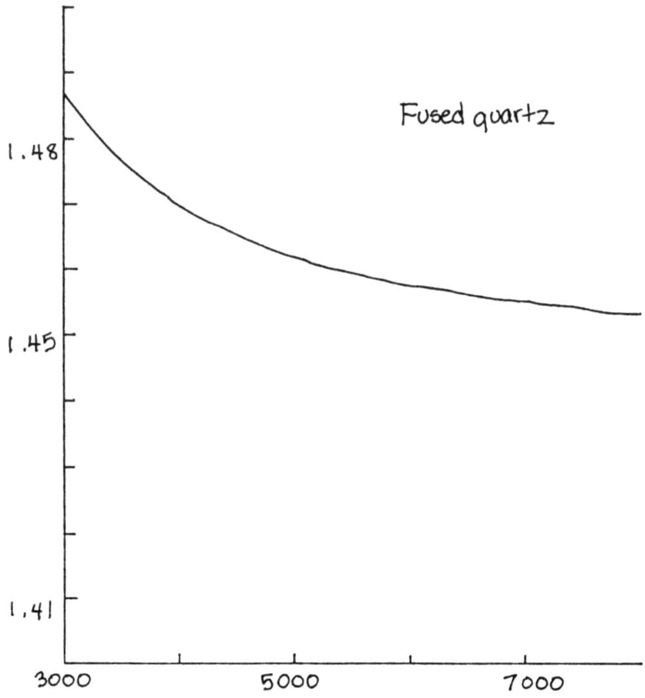

6. *Comparison of dielectrics and metals.* For the dielectric case, we have

$$n^2 - K^2 = 1 + \frac{\omega_p^2(\omega_0^2 - \omega^2)}{(\omega_0^2 - \omega^2)^2 + \gamma^2\omega^2} \tag{7.76}$$

$$2nK = \frac{\gamma\omega\omega_p^2}{(\omega_0^2 - \omega^2)^2 + \gamma^2\omega^2} \tag{7.77}$$

For the conducting case, we have

$$\bar{n}^2 - \bar{K}^2 = 1 - \frac{\omega_p^2}{\omega^2 + \gamma^2} \tag{7.78}$$

$$2\bar{n}\bar{K} = \frac{\gamma}{\omega}\left(\frac{\omega_p^2}{\omega^2 + \gamma^2}\right) \tag{7.79}$$

a. In which particular situation will the formula for $2nK$ approach the formula for $2\bar{n}\bar{K}$?
b. Show that for this same case, $n^2 - K^2$ approaches $\bar{n}^2 - \bar{K}^2$.
c. Show that (7.78) and (7.79) may be satisfied by a common value of $\bar{n} = \bar{K}$. Show that we get $\bar{K} = \bar{n} = \sqrt{\gamma/2\omega}$.
d. Show that $\bar{n} = \sqrt{\gamma/2\omega_p}$ if we assume that $\omega = \omega_p$ and $\bar{n} = \bar{K}$.

Solution

(a)

$$\omega_o = 0$$

(b)

$$2\,n\,K = \frac{\gamma\,\omega\,\omega_p^2}{\omega^4 + \gamma^2\omega^2} = \frac{\gamma}{\omega}\;\frac{\omega_p^2}{\omega^2 + \gamma^2} = 2\,\bar{n}\,\bar{K}$$

(c)

$$\frac{\omega_p^2}{\omega^2+\gamma^2} = 1$$

$$2\bar{n}^2 = \frac{\gamma}{\omega}\;\frac{\omega_p^2}{\omega^2+\gamma^2} = \frac{\gamma}{\omega}$$

$$\bar{K} = \bar{n} = \sqrt{\frac{\gamma}{2\omega}}$$

(d)

$$0 = 1 - \frac{\omega_p^2}{\omega_p^2+\gamma^2}$$

$$2\bar{n}^2 = \frac{\gamma}{\omega_p}\;\frac{\omega_p^2}{\omega_p^2 + \gamma^2}$$

$$2\bar{n}^2 = \frac{\gamma}{\omega_p}\;(1) \qquad \bar{n}^2 = \frac{\gamma}{2\omega_p}$$

7. *Inverse matrix.* Calculate the inverse matrix of

$$\begin{pmatrix} -\omega^2 m + g & i\omega e B_X \\ -i\omega e B_X & -\omega^2 m + g \end{pmatrix}$$

<u>Solution</u>

To calculate the inverse matrix of

$$\begin{vmatrix} -\omega^2 m + g & i\omega e B_x \\ -i\omega e B_x & -\omega^2 m + g \end{vmatrix} = \begin{vmatrix} d & b \\ -b & d \end{vmatrix}$$

We proceed as follows:

$$(a_{ij}) = \begin{pmatrix} d & b \\ -b & d \end{pmatrix} \qquad a_{ji} = \begin{pmatrix} d & -b \\ b & d \end{pmatrix}$$

$$M_{ji} = \begin{pmatrix} d & b \\ -b & d \end{pmatrix} \qquad A_{ji} = \begin{pmatrix} d & -b \\ b & d \end{pmatrix}$$

$$\text{Det.} \quad d^2 + b^2$$

$$(a_{ij})^{-1} = \begin{pmatrix} \frac{d}{d^2+b^2} & -\frac{b}{d^2+b^2} \\ \frac{b}{d^2+b^2} & \frac{d}{d^2+b^2} \end{pmatrix}$$

Proof:

$$\begin{pmatrix} d & b \\ -b & d \end{pmatrix} \begin{pmatrix} \frac{d}{\text{Det}} & -\frac{b}{\text{Det}} \\ \frac{b}{\text{Det}} & \frac{d}{\text{Det}} \end{pmatrix} = \begin{pmatrix} 1 & 0 \\ 0 & 1 \end{pmatrix}$$

---

8 Frequency Resolution and Fourier Transform Spectroscopy

1. *Reflection grating.* The diffracted intensity for an amplitude grating at angle θ is given by

$$u^2 = u_0^2 \frac{\sin^2 \gamma(d/2)}{[\gamma(d/2)]^2} \cdot \frac{\sin^2 N\gamma(a/2)}{\sin^2 \gamma(a/2)}, \quad \gamma = \frac{2\pi}{\lambda} \sin\theta$$

If we want to use such a grating in reflection, we obtain the same formula for the diffracted intensity if the light is incident normal to the plane of the reflecting surfaces of width d.

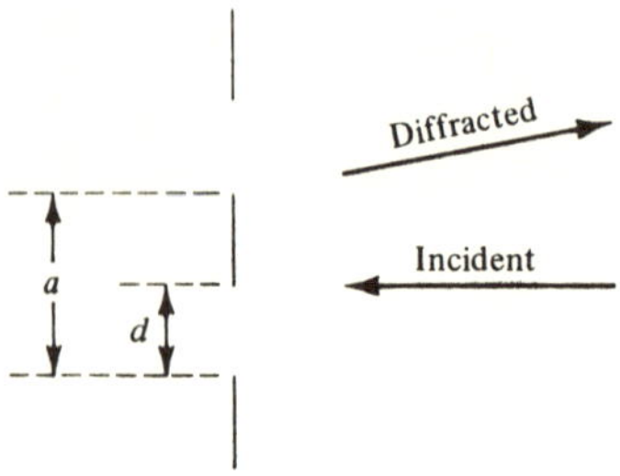

The maxima are obtained at

$$\frac{2\pi}{\lambda}\left(\frac{a}{2}\right)\sin\theta = m\pi$$

or

$$a\sin\theta = m\lambda$$

a. If $a = 10\ \mu$m, find the position of maxima for wavelength $\lambda = 2, \ldots, 7\ \mu$m in first and second orders.
b. At what angles θ do we observe overlapping of the second-order spectrum with the first-order spectrum?
c. What is the resolution of the grating if the length is 6 cm?
d. Give the value of the resolution in fifth and tenth orders.

<u>Solution</u>

(a)

		1 order	2 order
		$10\sin\theta = \lambda$	$10\sin\theta = 2\lambda$
$\lambda = 2$	$10\sin\theta = 2$	$\theta = 11.5°$	$\theta = 24°$
$\lambda = 3$		$\theta = 17°$	$\theta = 37°$
$\lambda = 4$		$\theta = 24°$	$\theta = 53°$
$\lambda = 5$		$\theta = 30°$	$\theta = 90°$
$\lambda = 6$		$\theta = 37°$	not possible
$\lambda = 7$		$\theta = 44°$	not possible

(b)

When $10\sin\theta = 4$ and $10\sin\theta = 2.2$ at $\theta = 24°$

$10\sin\theta = 6$ and $10\sin\theta = 2.3$ at $\theta = 37°$

(c)

$$\frac{\lambda}{\Delta\lambda} = mN \qquad N = \frac{6\times 10^4\mu m}{10\mu m} = 6\times 10^3$$

$$\frac{\lambda}{\Delta\lambda} = m\times 6\times 10^3$$

(d)

$$\frac{\lambda}{\Delta\lambda} = 30\times10^3 \quad \text{and} \quad \frac{\lambda}{\Delta\lambda} = 60\times10^3$$

5th order in 10th order

2. *Blazed grating in reflection.* A profile of the grating with steps as shown will direct the main intensity from the zero order to the first order.

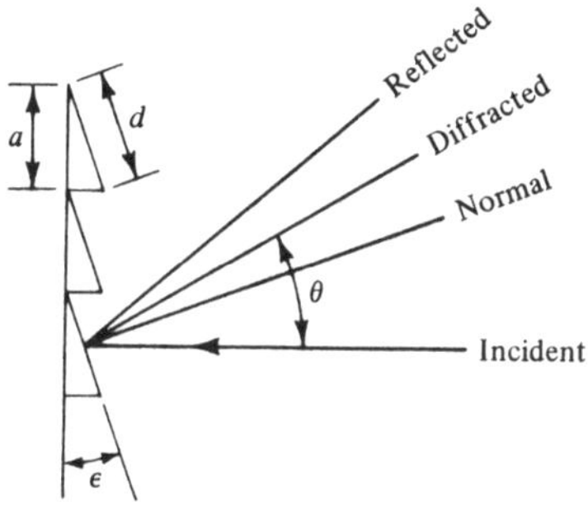

As in the transmission echelette grating, the interference factor remains the same, and the maxima are given by

$$a \sin\theta = m\lambda$$

The diffraction factor now has the main maximum at $\theta = 2\varepsilon$:

$$\frac{\sin^2[2\pi/\lambda(d/2)\sin(\theta - 2\varepsilon)]}{[2\pi/\lambda(d/2)\sin(\theta - 2\varepsilon)]^2}$$

If we choose $\theta = 2\varepsilon$ in the interference factor, we obtain the location of the maximum

$$a \sin 2\varepsilon = m\lambda$$

a. For $a = d = 10\ \mu$m and $\varepsilon = 15°$, find all the wavelengths for orders $m = 1, \ldots, 5$ that appear at the 30° angle.
b. Find the wavelength range for the maxima in first order for 15° around the 30° angle.
c. The intensity distribution in the range 15° around the 30° angle is given by the envelope obtained from the diffraction factor. Plot the envelope for the range $\lambda = 3, \ldots, 7\ \mu$m in steps of 1 μm.

Solution

(a)

$$a \sin 30 = m\lambda$$

$$10 \sin 30 = m\lambda$$

$$5\mu = m\lambda$$

m = 1	$\lambda = 5\mu$
m = 2	$\lambda = 2.5\mu$
m = 3	$\lambda = 1.66\mu$
m = 4	$\lambda = 1.25\mu$
m = 5	$\lambda = 1\mu$

(b)

$a = 10\mu m$, $\epsilon = 15°$, $2\epsilon = 30°$

$a \sin 15° = \lambda_1 \rightarrow 2.5\mu$

$a \sin 45 = \lambda_2 \rightarrow 7\mu$

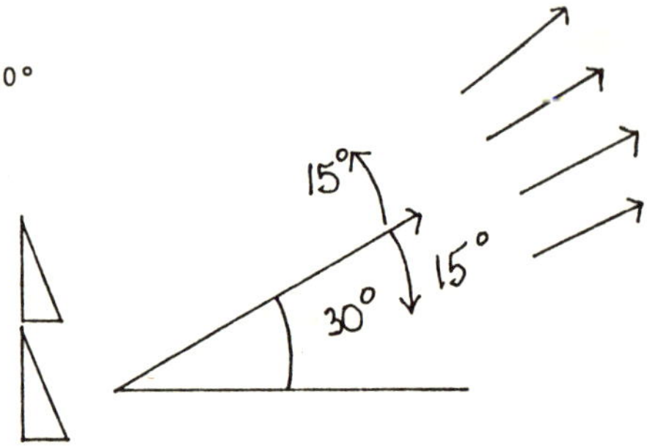

(c) $a = 10\mu m$, $2\epsilon = 30$, $d = 10\mu$

λ	$\frac{\lambda}{a} = \sin\theta$ θ	$\theta - 2\epsilon$ $\theta - 30$	$\sin(\theta-2\epsilon)$	$\frac{d}{\lambda}\pi$
3	17.5	-12.5	- .22	10.5
4	23.6	6.4	- .11	7.86
5	30.0	0	0	6.28
6	36.9	6.9	.12	5.23
7	44.5	14.5	.25	4.49

$\frac{d}{\lambda}\pi \sin(\theta-2\epsilon)$	$\sin(\frac{d}{\lambda}\pi\sin(\theta-2\epsilon)$	$\frac{\sin^2(\frac{d}{\lambda}\pi \sin(\theta-2\epsilon)}{(\frac{d}{\lambda}\pi \sin(\theta-2\epsilon)^2}$
-2.31	-0.74	.10
-0.86	-0.76	0.78
0	0	1
.63	0.59	.88
1.12	0.90	.65

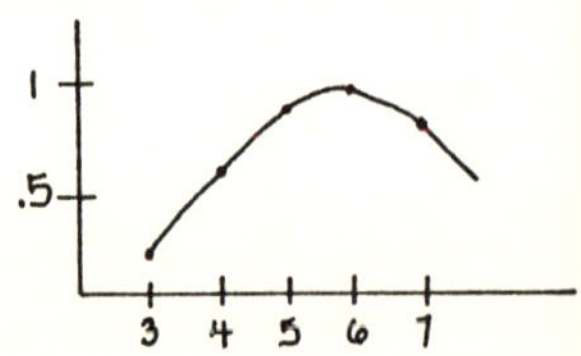

3. *Fabry-Perot etalon.* The transmitted intensity of a Fabry-Perot spectrometer is

$$I_t = \frac{1}{1 + g^2 \sin^2(2\pi D/\lambda)}, \quad g = \frac{2r}{1 - r^2}$$

To use the Fabry-Perot etalon as a spectrometer, we must vary the distance D. The wavelength λ for which resonance occurs is the transmitted wavelength. Assume that a filter passes wavelengths from 2 to 10 μm only.

a. What is the variation of D needed to cover this region?
b. At what wavelength do we observe an overlapping of the second order (of the shorter wavelength) on the first order? The region of no overlapping is called the free spectral range.
c. Assume that $r = 0.99$ and $r = 0.999$, and calculate $\Delta\lambda$ for $\lambda = 5$ μm for the tenth and one hundredth orders.

Solution

(a)

$$\sin 2\pi \frac{D_1}{\lambda_1} = \pi \qquad \sin 2\pi \frac{D_2}{\lambda_2} = \pi$$

$$D_1 = \frac{\lambda_1}{2} \qquad D_2 = \frac{\lambda_2}{2}$$

$$D_1 = 1\mu \qquad D_2 = 5\mu$$

(b)

At D corresponding to $2\lambda_1 = \lambda_2$

$$D = 2\mu\text{m}$$

(c)

$$.99 \rightarrow g = \left(\frac{2r}{1-r^2}\right) = \frac{1.98}{0.0199} = 99.5$$

$$.999 \rightarrow g = \quad = \frac{1.998}{.001999} = 999.5$$

$$\Delta\lambda = \frac{2\lambda}{\pi g m}$$

For			
	.99	m = 10	$\Delta\lambda$ = .0032
	.99	m = 100	$\Delta\lambda$ = .00032
	.999	m = 10	$\Delta\lambda$ = .00032
	.999	m = 100	$\Delta\lambda$ = .000032

4. *Higher orders of a Fabry-Perot etalon.* Light of continuous wavelength of $\lambda = 4500$ Å to $\lambda = 5500$ Å is incident on a Fabry-Perot etalon of thickness $D = 5 \cdot 10^{-4}$ cm. At what wavelength and what order do we observe resonance, that is, do we see an interference fringe?

Solution

The resonance condition is

$$D = m\,\frac{\lambda}{2}$$

D in Å is 5×10^{-4} cm $= 5 \times 10^{-4} \times 10^{8}$ Å

$$= 5 \times 10^{4}\text{ Å} = 50{,}000$$

$$\lambda = \frac{2 \times 50{,}000}{m}\text{ Å} = \frac{100{,}000}{m}\text{ Å}$$

m = 22	→	4545 Å
m = 21	→	4762 Å
m = 20	→	5000 Å
m = 19	→	5263 Å

5. *Michelson interferometer.* The intensity of a Michelson Interferometer is given as

$$u^2 = 4A^2 \cos^2\left(\frac{2\pi}{\lambda}(x_2 - x_1)\right)$$

where $x = x_2 - x_1$ is the distance of the Michelson mirror from its symmetric position. The first-order maximum is obtained for wavelength λ at $x = \lambda/2$. By varying x, we may get first-order maxima for different λ.

a. Consider the wavelength range 2–10 μm. What variation of x is necessary to cover the range?
b. Find the free spectral range for the first order.
c. Give the resolution for $\lambda = 5\ \mu$m and $m = 10$; for $\lambda = 5\ \mu$m and $m = 100$.

Solution

(a)

$$\lambda = 2 \rightarrow 10\mu$$

$$x \rightarrow 1\mu \rightarrow 5\mu$$

(b)

Overlapping $\quad 2\lambda_2 = \lambda_1$

$$\lambda_{1\text{ highest}} = 4\mu$$

(c)

$$\Delta\lambda = \frac{\lambda}{2m} = \frac{5}{20} = .25\mu m$$

$$\Delta\lambda = \frac{5}{200} = .025\mu m$$

6. *Sodium D lines.* The two sodium lines, called D lines, are at $\lambda = 5896$ Å and $\lambda = 5890$ Å.
   a. Apply the formula for the resolution of a grating, $\lambda/\Delta\lambda = mN$, and determine the number of lines N to be used for the resolution of the sodium lines if $m = 3$.
   b. What value B for the base of a prism must we use to accomplish this resolution of the sodium D lines? For glass, at $\lambda = 5461$ Å, $n = 1.460$, and at $\lambda = 5893$ Å, $n = 1.458$.
   c. If we use a Fabry-Perot etalon, the resolving power depends on the reflectivity of the plates. Assume that $r = 0.9$. What order of the Fabry-Perot etalon must be used to resolve the sodium D lines?

Solution

(a)

$$\lambda = \frac{1}{2}(5896 + 5890)\ \text{Å} = 5893\ \text{Å}$$

$$\Delta\lambda = 6\ \text{Å}$$

$$m = 3$$

$$N = \frac{5893\ \text{Å}}{3 \times 6\ \text{Å}} = 327 \text{ lines}$$

(b)

$$\frac{\lambda}{\Delta\lambda} = B\frac{dn}{d\lambda} = \frac{5896}{6} \cong 983$$

For glass: $$\frac{dn}{d\lambda} = \frac{\Delta n}{\Delta\lambda} = \frac{0.002}{432}\ \frac{1}{\text{Å}}$$

$$B \cdot \frac{0.002}{432} = 983\ \text{Å}$$

$$B = 983 \times 432\ \frac{1}{0.002}\ \text{Å}$$

$$B = 2\cdot 10^{8}\ \text{Å} = 2\ \text{cm}$$

(c)

$$983 = \frac{\lambda}{\Delta\lambda} = m\pi\frac{r}{1-r^2} = 983 = m\pi\frac{0.9}{0.19} = m\cdot 15$$

$$m = \frac{982}{15} \simeq 65$$

7. *Free spectral range, overlapping orders.*
   a. Consider a grating of 60 lines per millimeter and two spectral lines of 4000 Å and 5000 Å. Calculate m_{highest} and the corresponding angle θ (if $m = 0$ corresponds to $\theta = 0$).
   b. For the Fabry-Perot etalon, we have the same formula for the highest order, $m_h = \lambda_1/\overline{\Delta\lambda}$, which in our case equals 4. For overlapping in the ring pattern, we have

$$\overline{\Delta\lambda} = \frac{\lambda_1^2}{2D\cos\theta}$$

We set $\cos\theta = 1$ and calculate $D = 8 \cdot 10^3$ Å. This value is for a Fabry-Perot unreasonably small. A Fabry-Perot etalon is used for small $\overline{\Delta\lambda}$ and large values of D. Calculate m_h and D for $\overline{\Delta\lambda} = 1$ Å and $\lambda_1 = 4000$ Å.

Solution

(a)

$$(\overline{\Delta\lambda})m_h = (\lambda_2 - \lambda_1)m_h = \lambda_1$$

$$1000 \cdot m_h = 4000$$

$$m_h = 4$$

The grating constant for 60 lines/mm is

$$a = \frac{1}{60} = 0.016\text{mm} = 0.0016\text{cm}$$

$$= 0.0016\text{x}10^8 \text{Å}$$

$$a = 16\text{x}10^4 \text{Å}$$

$$m_h\lambda = a \sin\theta$$

$$4\text{x}5000\text{Å} = 16\text{x}10^4 \text{Å} \sin\theta$$

$$\sin\theta = \frac{4\text{x}5000}{16 \cdot 10^4}$$

$$= \frac{2 \text{x} 10^4}{16\text{x}10^4} = .125$$

$$\theta = 7.2°$$

(b)

$$m_h = \frac{4000}{1} = 4000$$

$$D = \frac{\lambda^2}{2\overline{\Delta\lambda}} = \frac{16 \text{x}10^6}{2} = 8\text{x}10^6 \text{Å}$$

$$= 0.8\text{mm}$$

$$= 800\mu\text{m}$$

8. *Fourier transformation of a Fourier transformation.* For the Gaussian function $e^{-(a^2/2)\nu^2}$, show that the Fourier transformation of the Fourier transformation is the function itself. Use

$$\int_0^{\infty} e^{-c^2x^2}\cos bx\,dx = \frac{\sqrt{\pi}}{2c}e^{-b^2/4c^2}$$

Solution

We want to calculate

$$\int_{-\infty}^{+\infty} e^{-a^2\frac{\nu^2}{2}}\, e^{i2\pi\nu y}\,d\nu$$

The function $e^{-\frac{a^2\nu^2}{2}}$ is even, therefore we have

$$2\int_0^{\infty} e^{-\frac{a^2\nu^2}{2}}\cos 2\pi\nu y\,d\nu$$

We use the integral formula

$$\int_0^{\infty} e^{-c^2x^2}\cos bx\,dx = \frac{\sqrt{\pi}}{2c}\,e^{-\frac{b^2}{4c^2}}$$

We set $\nu = x$, $2\pi y = b$, $\frac{a^2}{2} = c^2$

and have

$$2\int_0^{\infty} e^{-\frac{a^2\nu 2}{2}}\cos 2\pi\nu y\,d\nu = 2\,\frac{\sqrt{\pi}}{2\frac{a}{\sqrt{2}}}\,e^{-\frac{4\pi^2y^2}{4\frac{a^2}{2}}}$$

$$= \frac{\sqrt{2\pi}}{a}\,e^{-\frac{2\pi^2y^2}{a^2}}$$

A second transformation is

$$2\int_0^{\infty}\frac{\sqrt{2\pi}}{a}\,e^{-\frac{2\pi^2y^2}{a^2}}\cos 2\pi\nu y\,dy =$$

We substitute into the formula

$$y = x,\quad 2\pi\nu = b,\quad \frac{2\pi^2}{a^2} = c^2$$

$$= 2\,\frac{\sqrt{2\pi}}{a}\,\frac{\sqrt{\pi}}{2\frac{\sqrt{2}}{a}\cdot\pi}\,e^{-\frac{4\pi^2\nu^2}{4\frac{2\pi^2}{a^2}}} = e^{-\frac{a^2\nu^2}{2}}$$

9. *Resolving power of a grating, Fourier transform spectrometer, and Fabry-Perot spectrometer.* In the section on Fourier transform spectroscopy, we mentioned that the resolution depends on the optical path difference of the two superimposed beams. For a grating, we can use an angle of incidence close to 90° and an angle of diffraction close to 90°.

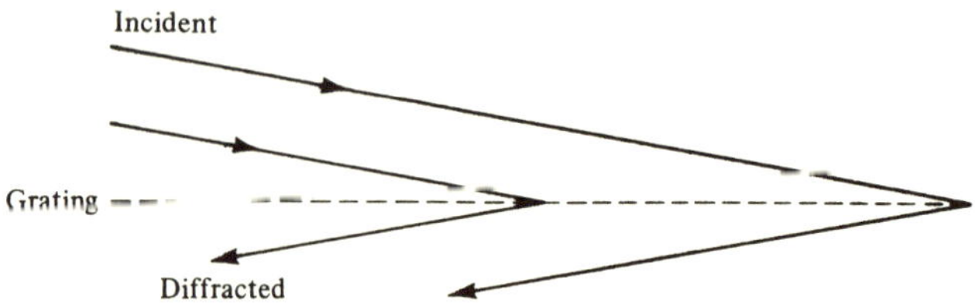

The path difference between the first and last wave is on the side of incidence equal to the length of the grating. For diffraction of the first order the path difference of the first and last wave is as well equal to the length of the grating. We have

$$L = Na$$

and

$$a(\sin\alpha_1 + \sin\alpha_2) = m\lambda$$

where a is the grating constant. The resolving power of the grating was given as

$$\frac{\lambda}{\Delta\lambda} = mN \quad \text{with} \quad N = \frac{L}{a} \quad \text{and} \quad 2a = m\lambda$$

and we have

$$\frac{\lambda}{\Delta\lambda} = m\frac{L}{a} = \frac{2a}{\lambda}\frac{L}{a} = \frac{2L}{\lambda}$$

$$\Delta\lambda = \frac{\lambda^2}{2L}$$

For the Fourier transform spectrometer, we have $\Delta\nu = 1/2L$ in 1/cm, where L is the maximum displacement of the Michelson mirror:

$$\lambda\nu = 1$$

$$\Delta\nu = \Delta\frac{1}{\lambda} = \left|\frac{\Delta\lambda}{\lambda^2}\right|, \quad \frac{\Delta\lambda}{\lambda^2} = \frac{1}{2L}, \quad \text{or} \quad \Delta\lambda = \frac{\lambda^2}{2L}$$

For the Fabry-Perot etalon with $m\lambda/2 = L$, where L is the thickness of the etalon,

$$\frac{\Delta\lambda}{\lambda} = \frac{2}{m\pi g} = \frac{2}{L(2/\lambda)\pi g} = \frac{\lambda}{L\pi g}$$

$$\Delta\lambda = \frac{\lambda^2}{\pi g L}$$

Whereas $\Delta\lambda/\lambda^2$ depends only on L for grating and Fourier transform spectra, we have the additional dependence of $\Delta\lambda/\lambda^2$ on $1/\pi g$ for the Fabry-Perot spectrum. The quantity $2/\pi g$ can be much smaller than 1. Therefore, the Fabry-Perot etalon has an advantage for resolution (in a limited small spectral region).

a. Calculate the resolution for $\lambda = 1\ \mu m$ and $L = 10$ cm for a grating.
b. For $\pi g = 2$, we would obtain the same result as in part (a) for the Fabry-Perot etalon. What is the resolution if we have $\pi g = 100$?

Solution

(a)
$$\Delta\lambda = \frac{1\,[\mu m]^2}{20\mathrm{x}10^4\ [\mu m]} = 5\mathrm{x}10^{-6}\,[\mu m]$$

(b)
$$\Delta\lambda = \frac{\lambda^2}{2L}\ \frac{2}{\pi g} = 5\mathrm{x}10^{-6}\,[\mu m]\ \frac{1}{5D}$$

$$= 10^{-7}\,[\mu m]$$

---

9 Spatial Resolution

1. *Mount Palomar telescope.* Calculate the resolution of a telescope with a mirror diameter of 5.1 m, assuming light with a wavelength of 5500 Å. Give the result in radians, degrees, and seconds.
Ans.
1.3×10^{-7} rad, 0.74×10^{-5} degrees, or 0.0268 s

Solution

$$\gamma = 1.22\ \frac{\lambda}{2a} = 1.22\ \frac{0.55 \times 10^{-4}\mathrm{cm}}{510\ \mathrm{cm}}$$

$$= 0.0013 \times 10^{-4}\ \mathrm{rad}$$

$$= 1.3 \times 10^{-7}\ \mathrm{rad}$$

γ in degrees : 0.74×10^{-5} degrees

γ in sec : 0.0268 sec

2. *Resolution of the eye.* Calculate the resolution for the pupil of the eye, given radii of $a = 0.25$, 0.5, 1, 2, and 4 mm, and assume a wavelength of $\lambda = 5500$ Å. The formula to use is

$$\gamma = 1.22\frac{\lambda/n}{2a}$$

where n is the refractive index of the eye. Assume that $n = 1.33$, which is close to the refractive index of water in the visible region. Give the results in radians and minutes.

Solution

$$\gamma = 1.22 \frac{0.55 \times 10^{-3} \text{mm}}{1.33 \times 2a}$$

in mm	in rad
a = 0.25	$\gamma = 1 \times 10^{-3}$ rad
a = 0.5	$\gamma = 0.50 \times 10^{-3}$ rad
a = 1	$\gamma = 0.25 \times 10^{-3}$ rad
a = 2	$\gamma = 0.125 \times 10^{-3}$ rad
a = 4	$\gamma = 0.063 \times 10^{-3}$ rad

in minutes

a = 0.25	$\gamma =$ 3.4 min
a = 0.5	$\gamma =$ 1.7 min
a = 1	$\gamma =$ 0.85 min
a = 2	$\gamma =$ 0.43 min
a = 4	$\gamma =$ 0.21 min = 12.6 sec

3. *Resolution of the microscope*
   a. Calculate the numerical aperture N.A. for $n = 1.55$ and $\alpha = 75°$.
   b. Calculate the size of an object in millimeters resolved by a microscope using the N.A. of part (a) and $\lambda = 5500$ Å.
   c. If we could use light of 3000 Å, how much would the resolution improve?
   d. The rule of thumb for the microscope is

$$Y_{\min} = 0.61 \cdot \frac{\lambda}{\text{N.A.}} = \frac{0.3}{\text{N.A.}} \mu\text{m}$$

for λ approximately in the middle of the visible region. The N.A. is given on the objective for nonimmersing liquids. Calculate $Y_{\min}$ for typical cases, such as N.A. = 0.25 and N.A. = 0.45.

Solution

(a)

$$\text{N.A.} = 1.55 \sin 75 = 1.5$$

(b)

$$Y_{\min} = .61 \frac{.55 \times 10^{-3} \text{ mm}}{1.49}$$

$$= .22 \times 10^{-3} \text{ mm} = .22 \mu\text{m}$$

(c)

$$Y_{\min} = .61 \frac{.30 \times 10^{-3}}{1.49} = 0.12 \mu\text{m}$$

About 2 times better.

(d)

$$Y_{min} = \frac{0.3}{0.25} = 1.2\mu m$$

$$Y_{min} = \frac{0.3}{0.45} = .67\mu m$$

---

10 Phase-Sensitive Imaging

1. *Fourier transformation, Gaussian function.* Functions for which $f(x) = f(-x)$ are even functions and symmetric with respect to $x = 0$. For such functions, the cosine-Fourier transform integral is all we need to use. We see this from

$$\int_{-\infty}^{+\infty} f(x)[\cos(xy) - i\sin(xy)]\,dy$$
$$= \int_{-\infty}^{+\infty} f(x)\cos(xy) - i\int_{-\infty}^{+\infty} f(x)\sin(xy)\,dy$$

a. Show that the second integral is equal to zero.
b. The Gaussian function is defined as $f(x) = e^{-a^2x^2}$. Calculate the Fourier transform of $f(x)$. Does it suffice to use only the cosine transformation?
c. The inverse Fourier transform was defined as

$$f(x) = \frac{1}{2\pi}\int_{-\infty}^{+\infty} F(y)e^{ixy}\,dy$$

Take the result from part (b), put it into the inverse Fourier transformation integral, and obtain back $f(x) = e^{-a^2x^2}$. We see for the Gaussian function, we can show in a simple way that the Fourier transformation of the Fourier transform is the original function.

Solution

(a)

$$\int_{-\infty}^{0} f(x)\sin xy\,dx + \underbrace{\int_{0}^{+\infty} f(x)\sin xy\,dx}$$

for $x \to -x$ we get

$$\int_{-\infty}^{0} f(x)\sin xy\,dx + \underbrace{\int_{0}^{-\infty} f(x)(-\sin xy)(-dx)}_{\int_{0}^{-\infty} f(x)\sin xy\,dx}$$

$$\int_{-\infty}^{0} f(x)\sin xy\,dx - \int_{-\infty}^{0} f(x)\sin xy\,dx = 0$$

(b) Only cos - transformation is necessary since $e^{-a^2x^2}$ is symmetric with respect to $x = 0$

$$F(y) = \int_{-\infty}^{+\infty} e^{-a^2x^2} \cos xy \, dx$$

We use the integral formula $\int_0^\infty e^{-a^2x^2} \cos bx \, dx = \frac{\sqrt{\pi}}{2a} e^{-\frac{b^2}{4a^2}}$

We have $F(y) = 2\int_0^\infty e^{-a^2x^2} \cos yx \, dx = 2\frac{\sqrt{\pi}}{2a} e^{-\frac{y^2}{4a^2}}$

$$F(y) = \frac{\sqrt{\pi}}{a} e^{-\frac{y^2}{4a^2}}$$

(c)

$$f(x) = \frac{1}{2\pi} \int_{-\infty}^{+\infty} \frac{\sqrt{\pi}}{a} e^{-\frac{y^2}{4a^2}} \cos xy dy$$

$$= \frac{\sqrt{\pi}}{2\pi a} 2 \int_0^\infty e^{-\frac{y^2}{4a^2}} \cos xy dy$$

call $\frac{1}{4a^2} = c^2$, $\int_0^\infty e^{-c^2y^2} \cos xy dy = \frac{\sqrt{\pi}}{2c} e^{-\frac{x^2}{4c^2}}$

$$f(x) = \frac{\sqrt{\pi}\,2}{2\pi a} \frac{\sqrt{\pi}}{2\frac{1}{2a}} e^{-\frac{x^2}{4\frac{1}{4a^2}}}$$

$$= e^{-a^2x^2}$$

2. Fourier transform: box. In Chapter 3, we calculated the amplitude diffraction pattern of a slit for Fraunhofer diffraction. The diffraction pattern may also be obtained as the Fourier transformation of the slit function P(x). We may use

$$f(x) = \frac{1}{2\pi}\int_{-\infty}^{+\infty} F(y)e^{ixy}\,dy, \quad F(y) = \int_{-\infty}^{+\infty} f(x)e^{-iyx}\,dx$$

where

$$f(x) = \begin{cases} 1 & \text{for} \quad -a \text{ to } +a \\ \text{otherwise } 0 \end{cases}$$

and the Fourier transformation yields

$$F(y) = 2a\frac{\sin(ay)}{ay}$$

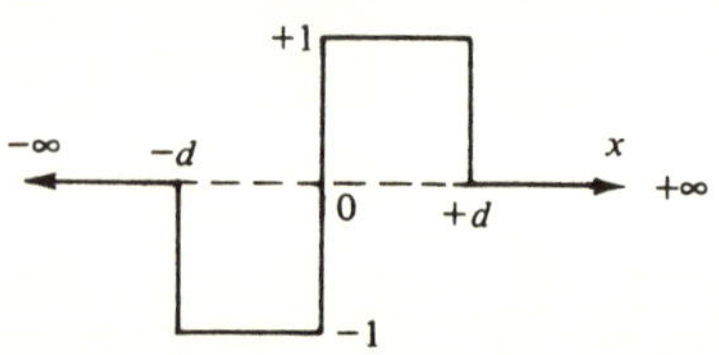

Calculate the Fourier transform of the function

$$\overline{f(x)} = \begin{cases} 0 & \text{for} \quad -\infty \text{ to } -d \\ -1 & \text{for} \quad -d \text{ to } 0 \\ +1 & \text{for} \quad 0 \text{ to } +d \\ 0 & \text{for} \quad d \text{ to } \infty \end{cases}$$

Make a sketch of the intensity diffraction pattern if $\overline{f(x)}$ is the aperture.

<u>Solution</u>

(a)

$$F(y) = \int_{-\infty}^{+\infty} \overline{f(x)} \; e^{-iyx} \, dx$$

$$= + \int_{-d}^{0} (-1) e^{-iyx} \, dx + \int_{0}^{+d} e^{-iyx} \, dx \qquad \frac{e^{-iyx}}{iy} \Big|_{-d}^{0} + \frac{e^{-iyx}}{-iy} \Big|_{0}^{d}$$

$$= + \frac{1}{iy} + \frac{e^{+iyd}}{-iy} + \frac{e^{-iyd}}{-iy} + \frac{1}{iy}$$

$$= + \frac{2}{iy} - \frac{e^{iyd} + e^{-iyd}}{iy}$$

$$= + \frac{2}{iy} - \frac{2}{iy} \cos yd = - \frac{2i}{y} + \frac{2i}{y} \cos yd$$

$$= - \frac{2i}{y} (1 - \cos yd) = - \frac{2i}{y} \, 2 \sin^2 \frac{yd}{2}$$

$$F(y) = - 4iyd^2 \, \frac{\sin^2 \frac{yd}{2}}{(yd)^2} = - iyd^2 \left(\frac{\sin \frac{yd}{2}}{\frac{yd}{2}} \right)^2$$

$$F(y) \cdot F(y)^* = y^2 d^4 \left(\frac{\sin \frac{yd}{2}}{\frac{yd}{2}} \right)^4$$

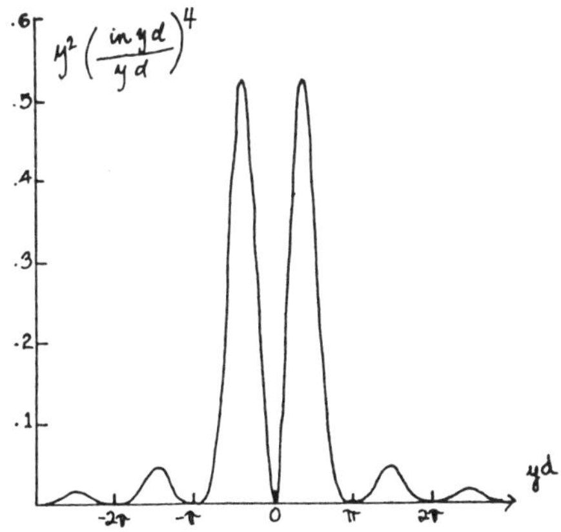

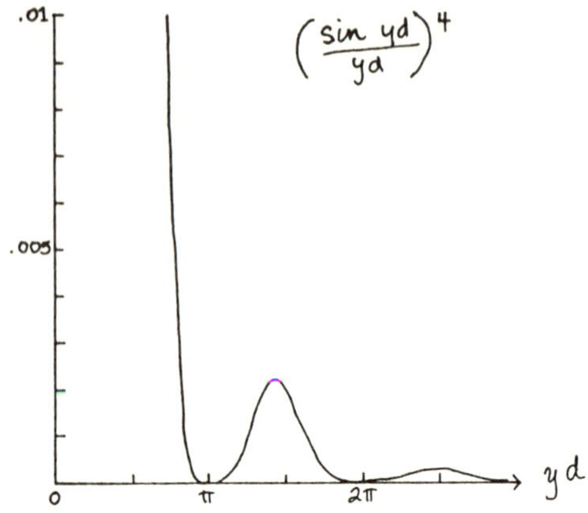

3. *Fourier transform: periodic function.* Consider the periodic function

$$f(x) = \begin{cases} \cos(y_0 x) & \text{for} \quad -a \text{ to } +a \\ \text{otherwise } 0 \end{cases}$$

Calculate the Fourier transform and make a sketch. The Fourier transformation may be interpreted as the distribution of spatial frequencies of the periodic pulse function.

Solution

(a)

$$F(y) = \int_{-\infty}^{+\infty} f(x) \cos xy dx = 2 \int_0^{\infty} f(x) \cos xy dx$$

since f(x) is symmetric $\quad f(x) = \cos y_o x : -a \leq x \leq a$

$$\frac{1}{2} (\cos(\alpha+\beta) + \cos(\alpha-\beta)) = \cos\alpha \cos\beta$$

$$2 \int_0^{\infty} [\cos y_o x \cos yx] dx = \int_0^a [\cos(y_o x+yx) + \cos(y_o x-yx)] dx$$

$$F(y) = \left.\frac{\sin(y_o+y)x}{y_o+y}\right|_0^a + \left.\frac{\sin(y_o-y)x}{y_o-y}\right|_0^a$$

$$= a[\text{sinc } (y_o+y)a + \text{sinc } (y_o-y)a]$$

These are the superposition of two sinc functions ($\text{sinc } x = \frac{\sin x}{x}$) one is centered at the position at $y = y_o$, the other at $y = -y_o$

4. *Fourier series.* A periodic function can be represented by the superposition of cosine and sine functions with different frequencies and amplitudes. For simplicity, we consider only odd functions, that is, $f(x) = -f(-x)$ with respect to $x = 0$. For

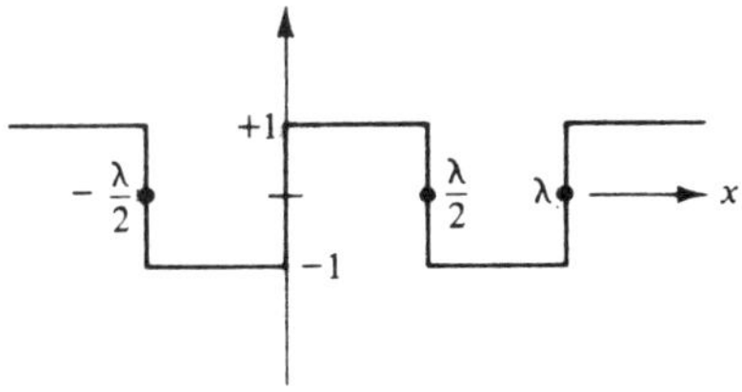

example, consider the following periodic function defined in the interval 0 to λ:

$$f(x) = \begin{cases} +1 & \text{for} \quad 0 < x < \dfrac{\lambda}{2} \\ -1 & \text{for} \quad \dfrac{\lambda}{2} < x < \lambda \end{cases}$$

We start by determining the sum of sine functions that can represent this function:

$$f(x) = \sum_{m=1}^{\infty} B_m \sin\left(2\pi m \frac{x}{\lambda}\right)$$
$$= \sum_{m=1}^{\infty} B_m \sin(mkx) \quad \text{where} \quad k = \frac{2\pi}{\lambda}$$

a. Calculate

$$\frac{2}{\lambda}\int_0^{\lambda} f(x)\sin(nkx)\,dx$$

by introducing

$$f(x) = \sum_{m=1}^{\infty} B_m \sin mkx$$

into the integral and show that only the term with $n = m$ is not equal to zero and that we get

$$B_m = \frac{2}{\lambda}\int_0^{\lambda} f(x)\sin mkx\,dx$$

b. Introduce the definition of $f(x)$ given in the preceding information and calculate B_m.
c. Calculate B_1 to B_5 and give the Fourier series.
d. Add the first three terms to get an idea of how fast the approximation works.
e. In the Fourier transformation integral, we obtained the spatial frequencies from the Fourier transformation of the space functions. Here we can read off the frequencies and their weights (amplitudes) from the series representation. Draw the frequency spectrum.

Solution

(a)

$$\frac{2}{\lambda}\int_0^{\lambda}\sum_{m=1}^{\infty} B_m \sin(m\,k\,x)\sin(n\,k\,x)\,d\,x$$

$$= \frac{2}{\lambda} \sum_{m=1}^{\infty} B_m \underbrace{\int_0^{\lambda} \sin(m\ k\ x)\sin(n\ k\ x)\,d\,x}_{\text{all intergrals for } m\neq n \text{ are zero}}$$

$$= \frac{2}{\lambda} B_n \underbrace{\int_0^{\lambda} \sin^2(n\ k\ x)\,d\,x}_{\frac{\lambda}{2}} = B_n$$

(b)

$$B_m = \frac{2}{\lambda} \int_0^{\frac{\lambda}{2}} \sin m\ k\ x\ d\ x + \frac{2}{\lambda} \int_{\frac{\lambda}{2}}^{\lambda} (-1) \sin m\,k\ x\ d\,x$$

$$= \frac{2}{\lambda} \left| - \frac{\cos\ m\ k\ x}{m\ k} \right|_0^{\frac{\lambda}{2}} - \frac{2}{\lambda} \left| - \frac{\cos\ m\ k\ x}{m\ k} \right|_{\frac{\lambda}{2}}^{\lambda}$$

$$= \frac{-2}{\lambda} \left(\frac{\cos\frac{2\pi}{\lambda}\ m\ \frac{\lambda}{2} - 1}{m\ k}\right) + \frac{2}{\lambda} \left(\frac{\cos\ 2\pi m - \cos\frac{2\pi}{\lambda}\ m\ \frac{\lambda}{2}}{m\ k}\right)$$

$$= -\frac{2}{\lambda m k} (\cos\pi m - 1 - 1 + \cos\pi m)$$

$$= \frac{-4}{\lambda m\frac{2\pi}{\lambda}} (\ \cos\pi m - 1) = \frac{2}{\pi m} (\ 1 - \cos\pi m)$$

(c)

$$B_1 = \frac{2}{\pi} (1 - \cos\pi) = \frac{4}{\pi} \quad , \quad B_2 = 0$$

$$B_3 = \frac{4}{3\pi} \quad , \quad B_4 = 0 \quad , \quad B_5 = \frac{4}{5\pi}$$

$$f(x) = \frac{4}{\pi} \sin\ 2\pi\ \frac{x}{\lambda} + \frac{4}{3\pi}\ \sin\ 2\pi\frac{x}{(\frac{\lambda}{3})}$$

$$+ \frac{4}{5\pi} \sin\ 2\pi\frac{x}{(\frac{\lambda}{5})}$$

(d) See next page.

(e)

Amplitudes	Wavelength
$\frac{4}{\pi}$	λ
$\frac{1}{3}\ \frac{4}{\pi}$	$\frac{\lambda}{3}$
$\frac{1}{5}\ \frac{4}{\pi}$	$\frac{\lambda}{5}$

(d)

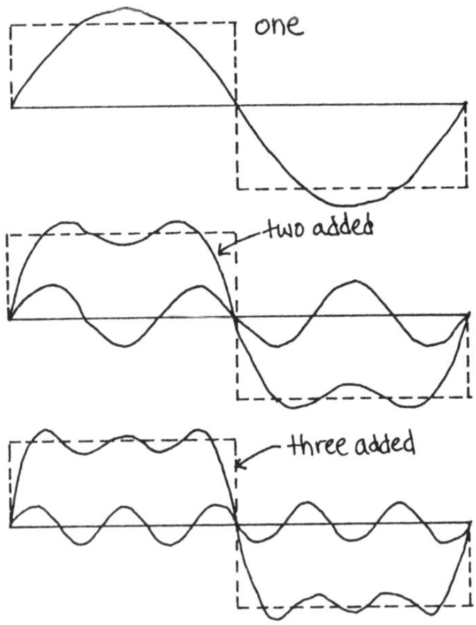

(e)

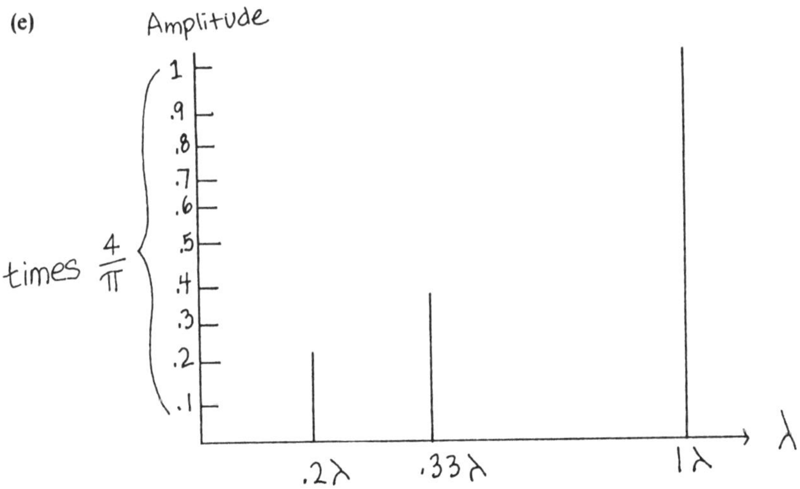

5. *The Dirac delta function and Fourier transformation.* The Dirac delta function is defined by

$$\delta(x) = \begin{cases} 0 & \text{for} \quad x \neq 0 \\ \infty & \text{for} \quad x = 0 \end{cases} \quad \text{and} \quad \int_{-\infty}^{+\infty} \delta(x)\,dx = 1$$

If it is used as a product with another function $f(x)$ in an integral from $-\infty$ to $+\infty$, one can show

$$\int_{-\infty}^{+\infty} \delta(x - x_0) f(x)\,dx = f(x_0)$$

a. Calculate the integral

$$\int_{-\infty}^{+\infty} \delta(x - x_0) f(x)\, dx \quad \text{for} \quad f(x) = e^{-ixy}$$

b. Can we say that the Fourier transform of $\delta(x - x_0)$ is $e^{-ix_0 y}$?

c. Calculate the Fourier transform of

$$A[\delta(x - x_0) + \delta(x + x_0)]$$

Show that we get as result $2A\cos(x_0 y)$. Compare with the interference from two narrow slits in chapter 2.

In Chapter 2, we found that interference patterns were obtained by a summation process. In Chapter 3, we used an integration process to derive the diffraction pattern. In part (c) of this problem we obtained formally the interference pattern of two source points by integration. We calculated the Fourier transform of two Dirac delta functions.

d. Calculate the Fourier transformation of

$$\sum_{n=0}^{n=+N-1} A\delta(x - nx_0)$$

In Chapter 2, we summed up

$$\sum_{n=0}^{n=N-1} e^{inx_0 y}$$

and obtained

$$\frac{\sin[N(x_0 y/2)]}{\sin(x_0 y/2)} e^{i(N-1)(x_0 y/2)}$$

If we square this expression, we have the following interference pattern observed from N (narrow sources):

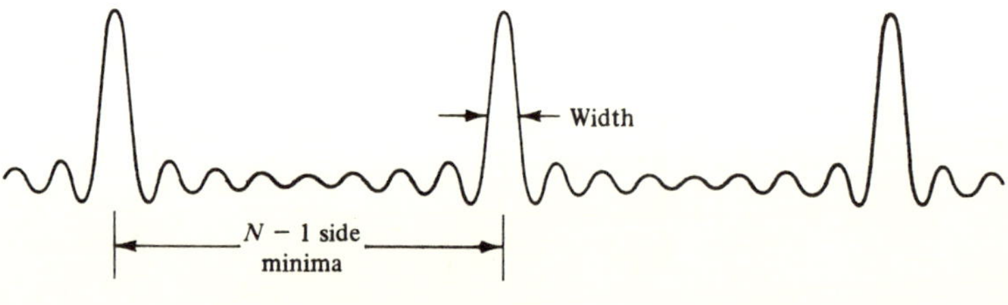

We found that for large values of N, the width of the main maxima becomes narrower, the number of side maxima increases, and their individual heights decrease. For $n \to \infty$, we expect that the Fourier transform

$$f(x) = \sum_{n=-\infty}^{n=+\infty} A\delta(x - nx_0)$$

is a sum of delta functions, and one can show that the Fourier transform is

$$F(y) = \sum_{n=-\infty}^{+\infty} \frac{2\pi A}{x_0} \delta\left(y - n\frac{2\pi}{x_0}\right)$$

e. If $f(x)$ is graphically expressed as

$$> |x_0| <$$

| | | | | | | } A

make a sketch of $F(y)$ and give the separation and heights of the "bars."

<u>Solution</u>

(a)

$$\int_{-\infty}^{+\infty} \delta\,(x-x_o)\, e^{-ixy}\, dx = e^{-ix_oy}$$

(b) yes

(c)

$$\int_{-\infty}^{+\infty} A[\,\delta(x - x_o) + \delta\; x + x_o)\,]\, e^{-ixy}\, dx$$

$$= A \int_{-\infty}^{+\infty} \delta\,(x - x_o)\; e^{-ixy}\, dx + A \int_{-\infty}^{+\infty} \delta(x + x_o)\; e^{-ixy}\, dx$$

$$= A\,[\,e^{-ix_oy} + e^{+ix_oy}\,] = 2A \cos x_oy$$

(d) For each term we get Ae^{-inx_oy}

The result is

$$A \sum_{n=0}^{n=N-1} e^{-inx_oy}$$

(e)

$$>|\,x_o\,|<\qquad\qquad\qquad\qquad >|\;\tfrac{2\pi}{x_o}\;|<$$

| | | | | | | }A $\xrightarrow{FT}$ | | | | | | | |}$\frac{2\pi A}{x_o}$

The reciprocity in distance and hight has been discussed in this chapter. The FT of a FT of an original grating structure was claimed to be the original grating structure.

6. *Phase grating*

a. Fill in all steps in the calculation of the phase grating and arrive at

$$u_p = 2e^{-ik(h/2)} \frac{-\sin[k(h/2] + \sin[k(d/2)(Y/X) - k(h/2)]}{k(Y/X)} \times \frac{\sin[kN(d/2)(Y/x)]}{\sin[k(d/2)(Y/X)]} e^{ikN(d/2)(Y/X)}$$

b. Show that we have

$$\frac{-\sin(kh/2)+\sin[\pi m+(kh/2)]}{(2m\pi/d)} = \begin{cases} \text{for} \quad m \text{ odd} & -\dfrac{d}{\pi m}\sin\left(\dfrac{kh}{2}\right) \\ \text{for} \quad m \text{ even but } \neq 0 & 0 \\ \text{for} \quad m=0 & \dfrac{d}{2}\cos\left(\dfrac{kh}{2}\right) \end{cases}$$

Solution

(a)

$$\int_0^{\frac{d}{2}} e^{iky\frac{Y}{X}}\,dy\cdot e^{-ikh} + \int_d^{d+\frac{d}{2}} e^{iky\frac{Y}{X}}\,dy\; e^{-ikh} \cdots\cdots \int_{(N-1)d}^{(N-1)d+\frac{d}{2}}$$

$$+\int_{\frac{d}{2}} e^{iky\frac{Y}{X}}\,dy \qquad + \int_{d+\frac{d}{2}}^{2d} e^{iky\frac{Y}{X}}\,dy \qquad + \quad \cdots\cdots \int_{Nd+\frac{d}{2}}^{Nd}$$

We use: $k\frac{Y}{X}y = \alpha y \quad kh = \beta$

$$\frac{d}{2}$$

$$= \frac{e^{i\alpha y}}{i\alpha}\Big|_0^{\frac{d}{2}}\cdot e^{-i\beta} + \frac{e^{i\alpha y}}{i\alpha}\Big|_d^{d+\frac{d}{2}} e^{-i\beta} + \cdots\cdots \frac{e^{i\alpha y}}{i\alpha}\Big|_{(N-1)d}^{(N-1)d+\frac{d}{2}} e^{-i\beta}$$

$$+\frac{e^{i\alpha y}}{i\alpha}\Big|_{\frac{d}{2}}^{d} \qquad + \frac{e^{i\alpha y}}{i\alpha}\Big|_{d+\frac{d}{2}}^{2d} \qquad + \cdots\cdots \frac{e^{i\alpha y}}{i\alpha}\Big|_{(N-1)d+\frac{d}{2}}^{Nd}$$

$$= +\frac{e^{i\alpha\frac{d}{2}}-1}{i\alpha}e^{-i\beta} + \frac{e^{i\alpha(d+\frac{d}{2})}-e^{i\alpha d}}{i\alpha}e^{-i\beta} + \cdots\cdots \frac{e^{i\alpha(N-1)d+\frac{d}{2}}-e^{i\alpha(N-1)d}}{i\alpha}e^{-i\beta}$$

$$+\frac{e^{i\alpha d}-e^{i\alpha\frac{d}{2}}}{i\alpha} \qquad + \frac{e^{i\alpha 2d}-e^{i\alpha(d+\frac{d}{2})}}{i\alpha} \qquad +\cdots\cdots \frac{e^{i\alpha Nd}-e^{i\alpha(N-1)d+\frac{d}{2}}}{i\alpha}$$

$$\frac{e^{i\alpha\frac{d}{2}}-1}{i\alpha}e^{-i\beta} + \frac{e^{i\alpha\frac{d}{2}}-1}{i\alpha}e^{-i\beta}e^{i\alpha d} + \cdots \frac{e^{i\alpha\frac{d}{2}}-1}{i\alpha}e^{-i\beta}e^{i\alpha(N-1)d}$$

$$\frac{e^{i\alpha\frac{d}{2}}-1}{i\alpha}e^{i\alpha\frac{d}{2}} + \frac{e^{i\alpha\frac{d}{2}}-1}{i\alpha}e^{i\alpha(d+\frac{d}{2})} + \cdots \frac{e^{i\alpha\frac{d}{2}}-1}{i\alpha} + e^{i\alpha(N-1)+\frac{d}{2}}$$

$$= \frac{e^{i\alpha\frac{d}{2}} - 1}{i\alpha} e^{-i\beta} (1 + e^{i\alpha d} + \cdots\cdots e^{i\alpha(N-1)d})$$

$$+ \frac{e^{i\alpha\frac{d}{2}} - 1}{i\alpha} e^{i\alpha\frac{d}{2}} \underbrace{(1 + e^{i\alpha d} + \cdots\cdots e^{i\alpha(N-1)d})}_{\frac{\sin N\frac{\alpha}{2}d}{\sin\frac{\alpha}{2}d} \; e^{i(N-1)\frac{\alpha}{2}d}}$$

$$= \frac{(e^{i\alpha\frac{d}{2}} - 1)}{i\alpha} (e^{-i\beta} + e^{i\alpha\frac{d}{2}}) \frac{\sin N\frac{\alpha}{2}d}{\sin\frac{\alpha}{2}d} e^{i(N-1)\frac{\alpha}{2}d}$$

$$= (e^{i\alpha\frac{d}{2} - i\beta} - e^{-i\beta} + e^{i\alpha d} - e^{i\alpha\frac{d}{2}}) \frac{\sin N\frac{\alpha}{2}}{i\alpha \sin\frac{\alpha}{2}} e^{i(N-1)\frac{\alpha}{2}d}$$

$$= (e^{-i\beta} - e^{-i\beta-\alpha\frac{d}{2}} + e^{i\alpha\frac{d}{2}} - 1) \left(\frac{\sin N\frac{\alpha}{2}d}{i\alpha \sin\frac{\alpha}{2}d} e^{iN\frac{\alpha}{2}d}\right)$$

$$= \left[e^{-i\frac{\beta}{2}}(e^{-i\frac{\beta}{2}} - e^{+i\frac{\beta}{2}}) + e^{-i\frac{\beta}{2}}(-e^{-i\frac{\beta}{2}-\alpha\frac{d}{2}} + e^{i\alpha\frac{d}{2}+i\frac{\beta}{2}})\right] \frac{\sin N\frac{\alpha}{2}d}{i\alpha\sin\frac{\alpha}{2}d} e^{iN\frac{\alpha}{2}d}$$

$$= e^{-i\frac{\beta}{2}} \left[-2i\sin\frac{\beta}{2} + 2i\sin(\alpha\frac{d}{2} + \frac{\beta}{2})\right] \left[\frac{\sin N\frac{\alpha}{2}d}{i\alpha\sin\frac{\alpha}{2}d} e^{iN\frac{\alpha}{2}d}\right]$$

$$= 2e^{-i\frac{\beta}{2}} \frac{-\sin\frac{\beta}{2} + \sin(\alpha\frac{d}{2} + \frac{\beta}{2})}{\alpha} \frac{\sin N\frac{\alpha}{2}d}{\sin\frac{\alpha}{2}d} e^{iN\frac{\alpha}{2}d}$$

$$= 2e^{-ik\frac{h}{2}} \frac{-\sin\frac{kh}{2} + \sin(k\frac{Y}{X}\frac{d}{2} + \frac{kh}{2})}{k\frac{Y}{X}} \frac{\sin Nk\frac{Y}{X}\frac{d}{2}}{\sin k\frac{Y}{X}\frac{d}{2}} e^{iNk\frac{Y}{X}\frac{d}{2}}$$

(b)

We have

$$-\sin\frac{kh}{2} + \sin(\pi m + \frac{kh}{2}) = \begin{cases} m \text{ odd } & -2\sin\frac{kh}{2} \\ m \text{ even } & 0 \end{cases}$$

because of

$$\sin(\pi + \alpha) = -\sin\alpha$$

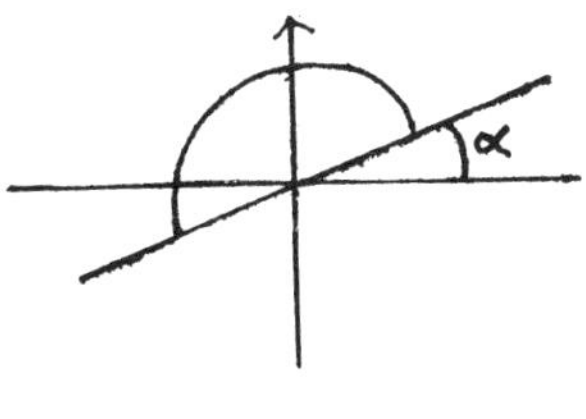

$m = 0$ differentiation with respect to m

$$\frac{-\sin k\,\frac{h}{2} + \sin\,(\pi\,m + k\,\frac{h}{2})}{\frac{2\pi m}{d}} \to \frac{\cos\,(\pi\,m + k\,\frac{h}{2})\,\pi}{\frac{2\pi}{d}}$$

$$m \to 0 \to \frac{d}{2}\,\cos\frac{kh}{2}$$

7. *Fresnel zone plate.* Derive the focal length of a Fresnel zone plate by assuming parallel light incident from the left. Check your result by comparison with the thin-lens equation for the zone plate when $R_0 \to \infty$.

Solution

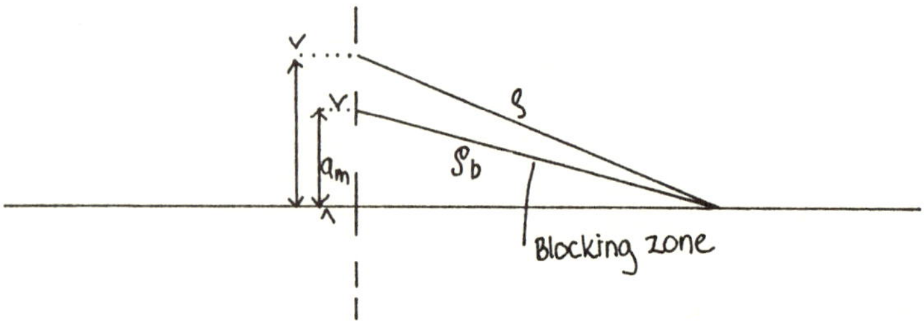

We have $a_m^2 + \rho_o^2 = \rho_m^2$ or $\rho_m = \sqrt{\rho_o^2 + a_m^2} \simeq \rho_o(1+\frac{1}{2}\frac{a_m^2}{\rho_o^2})$

For the blocking zone we have $\rho_b - \rho_o = m\,\frac{\lambda}{2}$

Combination; $m\,\frac{\lambda}{2} + \rho_o = \rho_o + \frac{1}{2}\,\frac{a_m^2}{\rho_o}$

or $$\frac{1}{\rho_o} = \frac{m\lambda}{a_m^2}$$

8. *Fresnel zone plate and chromatic aberration.* The focal length of the thin-lens equation for the zone plate (equations 10.18 and 10.19) depends on the wavelength. Parallel light incident on the plate will be focused to the focal point $f = \rho_0$. If the incident light has a different wavelength, we will obtain a different focal length. This corresponds to chromatic aberration of lenses (see Chapter 13). Calculate the difference in the focal length for light of wavelength $\lambda_1 = 0.5 \times 10^{-3}$ mm and $\lambda_2 = 0.55 \times 10^{-3}$ mm, where $\rho_0 = 1000$ mm.

Solution

$$\rho_{o1} = f_1 = \frac{a_m^2}{m(.5\cdot 10^{-3}[\mathrm{mm}])} \qquad \rho_{o2} = f_2 = \frac{a_{m'}^2}{m'(.55\cdot 10^{-3}[\mathrm{mm}])}$$

We have $\rho_o = f = \frac{a_m^2}{m\lambda}$. Since the difference in λ is very small

we take $\frac{a_m^2}{m} = \frac{a_{m'}^2}{m'}$ and have

$$f_1 x(.5 \times 10^{-3} [mm]) = f_2 x(.55 \times 10^{-3} [mm])$$

$$\frac{f_1}{f_2} = \frac{.55}{.5} = 1.1$$

$$\rho_{o1} = 1.1\ \rho_{o2} \quad , \quad \rho_{o1} = 1000mm \rightarrow \quad \rho_{o2} = 909mm$$

$$\Delta f = \Delta\rho = 91mm$$

---

11 Radiometry and Photometry

1. *Radiance of a star.* Calculate the radiance L for a star that has the irradiance of the sun and is so far away that its angular diameter could be determined by the Michelson stellar interferometer to be $\alpha = 0.05$ s.

<u>Solution</u>

$$L = \frac{E}{d\Omega} = \frac{1.35 x 10^3 \frac{W}{m^2}}{d\Omega}$$

$$d\Omega = \pi \sin^2(\alpha) \qquad \alpha = 0.05''$$

one has

$$\sin 1'' = 4.8 \times 10^{-6}\ rad$$

$$\sin 0.05'' = .24 \times 10^{-6}\ rad$$

$$\sin^2(0.05'') = .058 \times 10^{-12} (rad)^2$$

$$d\Omega = 0.18 \times 10^{-12} sr$$

$$L = \frac{1.35 x 10^3 \frac{W}{m^2}}{0.18 x 10^{-12} sr} = 7.5 x 10^{15} \frac{W}{(sr) m^2}$$

2. *Solid angle.* Show that the solid angle (in units of steradians) subtended by the right circular cone of half-angle α' is

$$\Omega = 4\pi \sin^2 \frac{\alpha'}{2}$$

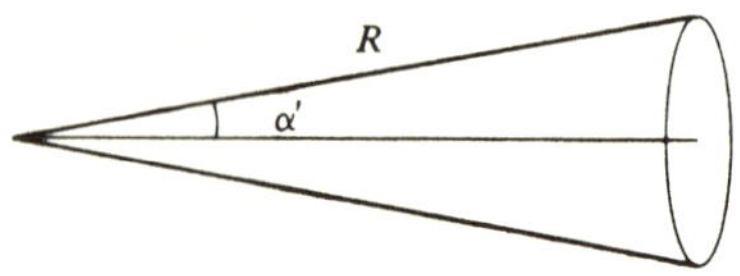

Solution

We consider the section of the sphere with α, $d\alpha$ and R

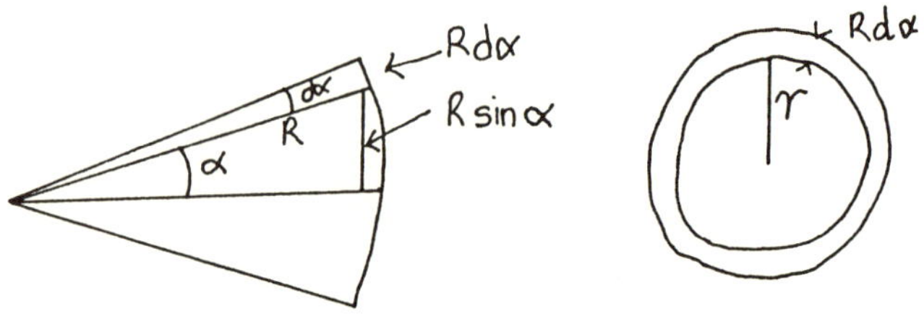

as indicated. The area of the circular section is $2\pi r\ Rd\alpha$.

Since $r = R \sin \alpha$ we have

$$d\Omega = 2\pi\ R \sin \alpha\ Rd\alpha$$

Integration from 0 to α' yields

$$\Omega = 2\pi R^2 \int_0^{\alpha'} \sin \alpha d\alpha = 2\pi R^2 \left| -\cos\alpha \right|_0^{\alpha'}$$

$$= 2\pi R^2 (-\cos \alpha' + 1)$$

Using $2 \sin^2 \frac{\alpha'}{2} = 1 - \cos \alpha'$ and setting $R = 1$

we have

$$\Omega = 4\pi \sin^2 \frac{\alpha'}{2}$$

3. *Radian and steradian for small angles*

a.
$$1 \text{ rad} = \frac{360°}{2\pi} = 57.3°$$

Calculate the number of radians for 1°; for 1 min; for 1 s.

b. How many degrees is 20 mrad if 1 mrad equals 10^{-3} rad?

c. For small angles, we have $d\Omega = \pi \sin^2 \alpha$ in steradians. Since α is small, we have $\sin \alpha \simeq \alpha$. If α is in degrees for $\sin \alpha$ (left side), it must be in radians for the right side. Check this out with a calculator for 1°, 1 min, and 1 s. Radians and steradians are dimensionless quantities. However, the "dimension" of the solid angle is $[\text{rad}]^2$ or [sr].

Solution

(a) $1° = 17.5 \times 10^{-3}$ rad

$1' = 2.9 \times 10^{-4}$ rad

$1'' = 4.8 \times 10^{-6}$ rad

(b) 1 rad => 57.3°

10^{-3} rad => 0.057°

20 millirad => = 1.15°

(c) sin 1° = 0.0175 rad   see solution A

sin 1' = 0.00029 rad

sin 1" = 0.0000048 rad

4. *Large emitter and small receiver area.* For the case of a large emitter area and a receiver with a small area da', do the inegration leading to

$$P = L\, da'\, \pi \sin^2 \beta'$$

Solution

From eq. (20) we have

$$dP = L\, 2\pi\, (x\, dx \cos\gamma) \frac{da' \cos\gamma}{(R')^2}$$

From Fig. 6 we have $R' = \dfrac{R}{\cos\gamma}$, $\quad x = R\tan\gamma$ and $dx = R \dfrac{1}{\cos^2\gamma} d\gamma$

$$dP - L\, 2\pi\, R\tan\gamma\, R \frac{1}{\cos^2\gamma} d\gamma \quad \cos\gamma \frac{da' \cos\gamma}{\dfrac{R^2}{\cos^2\gamma}}$$

$$= 2\pi\, L \sin\gamma \cos\gamma\, d\gamma\; da'$$

$$P = 2\pi\, L\, da' \int_0^{\beta'} \sin\gamma \cos\gamma\, d\gamma = 2\pi\, L\, da' \left| \frac{\sin^2\gamma}{2} \right|_0^{\beta'}$$

$$= L\, da' \,(\pi \sin^2 \beta')$$

5. *Power for Lambert and non-Lambert sources.* Compare the power P for Lambert and non-Lambert sources, assuming a small-area emitter and a large-area receiver, where the emitter area is da and the half-angle is α.

a. Show that the formulas are the same for small angles.

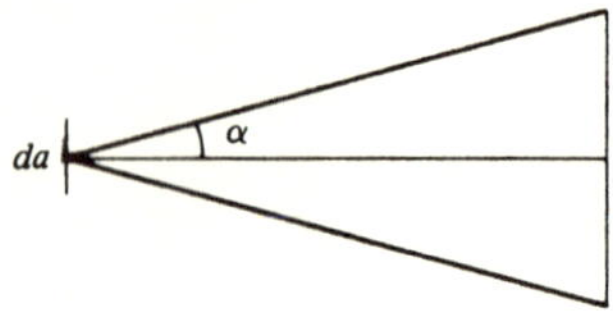

b. Give the ratio P_L/P_{NL} for $\alpha = 30°$.
c. Give the ratio P_L/P_{NL} for $\alpha = 60°$.

Solution

Lambert source

$$P_L = L\ da\ (\pi \sin^2\alpha)$$

$$P_{NL} = L\ da\ 4\pi \sin^2 \frac{\alpha}{2} \qquad \text{from 11.2}$$

(a) For small angles both are equal

$$P_L = L\ da\ \pi\ (\alpha)^2$$

$$P_{NL} = L\ da\ \pi\ 4\ (\tfrac{\alpha}{2})^2$$

(b) What is the ratio $\frac{P_L}{P_{NL}}$ for $\alpha = 30°$

$$P_L = L\ da\ \pi\ (0.250)$$

$$P_{NL} = L\ da\ \pi\ \underbrace{(4 \cdot 0.06)}_{0.268}$$

$$\frac{P_L}{P_{NL}} = \frac{0.250}{0.268} = .933$$

(c) What is the ratio $\frac{P_L}{P_{NL}}$ for $\alpha = 60°$

$$\frac{P_L}{P_{NL}} = \frac{\sin^2 60}{4 \sin^2 30} = \frac{.75}{1} = .75$$

6. *Irradiance of the sun with and without a magnifier.* For the sun, $E = 1.35 \times 10^3$ W/m^2. What is the irradiance if we use a magnifier of diameter 5 cm and focal length 5 cm?

Solution

We have for the irradiance with lens

$$E_1 = E_2 \frac{\sin^2\beta}{\sin^2\beta'}$$

where E_2 is the irradiance E of the Sun and β is the solid angle of the magnifier, β' of the Sun.

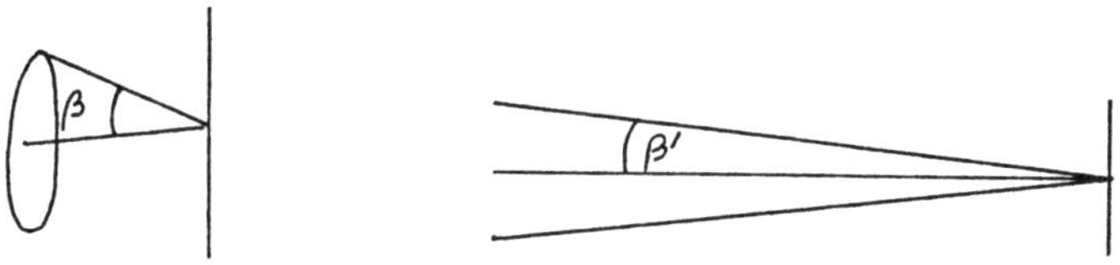

We calculate β: $\tan \beta = \frac{2.5}{5} \rightarrow \beta = 26.6°$

We have for β' = .25°, therefore

$$E_1 = 1.35\times10^3 \frac{W}{m^2} \frac{\sin^2 26.6}{\sin^2 .25}$$

$$= 1.35\times10^3 \frac{W}{m^2} \frac{.2}{19 \times 10^{-6}}$$

$$= 1.35\times10^3 \frac{.2}{.19 \times 10^{-4}} \frac{W}{m^2}$$

$$= 14.2 \times 10^6 \frac{W}{m^2}$$

7. *Sunlight and magnifier.* Suppose that sunlight is falling at normal incidence onto an area with a radius of 0.5 cm.

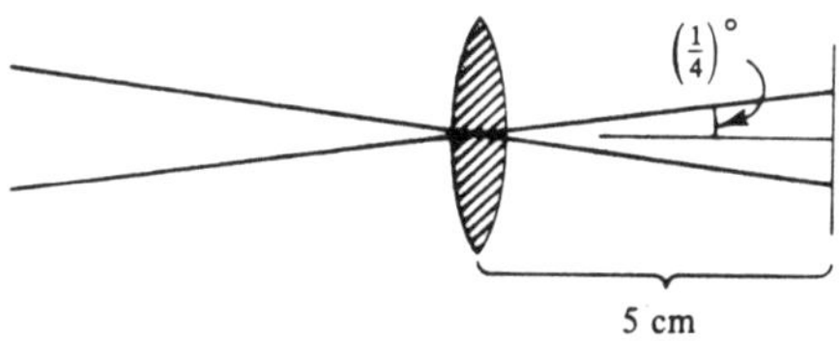

a. Calculate the incident power on the area.
b. A lens is placed before the area. Its radius is 0.5 cm and its focal length is 5 cm. Calculate the area *da'* of the image of the sun. Compare your answer with the one the paraxial theory would give for the area of the spot from chief rays from the sun.

Solution

(a) $P = L \cdot d\Omega \cdot da'$

$$L = 2.25\times10^7 \frac{W}{m^2 sr}, \quad d\Omega = 6\times10^{-5} sr, \quad da' = \pi\ (.5)^2 \times10^{-4} m^2$$

$$P = 2.25\times6\times(.79)\ 10^7 \times10^{-5} \times10^{-4} \frac{W}{m^2 sr}\ sr\ m^2$$

$$= 10.7 \times 10^{-2}\ W$$

(b) The solid angle from the image spot of the Sun to the lens is

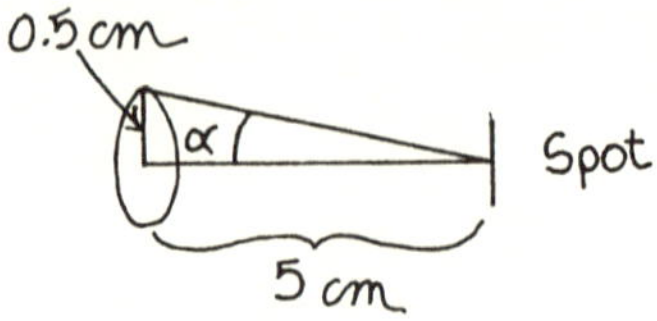

$$\tan \alpha = \frac{0.5}{5} = 0.1$$

$$\alpha = 5.7°$$

The solid angle is $\pi \sin^2 5.7° = 0.031$ sr

We have

$$P = L \cdot (\pi \sin^2 5.7) \cdot da'$$

$$da' = \frac{P}{L \cdot (\pi \sin^2 5.7)}$$

$$= \frac{10.7 \times 10^{-2} W}{2.25 \times 10^7 \frac{W}{m^2 sr} \cdot 0.031 \text{ sr}} =$$

$$= \frac{10.7 \times 10^{-2} \ m^2}{(2.25)(0.031) \times 10^{+7}} = 151 \times 10^{-9} \ m^2$$

$$= 15.1 \times 10^{-8} \ m^2$$

From paraxial theory:

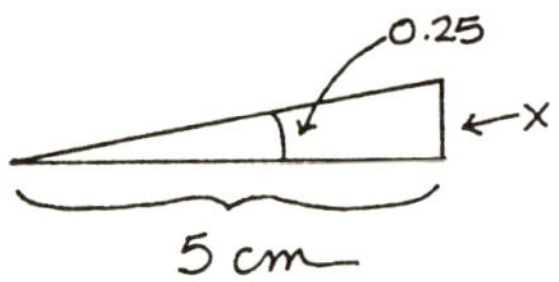

$$\frac{x}{5cm} = \tan 0.25°$$

$$x = 0.05m \ (0.0043) = 0.0002m$$

$$\pi \ x^2 = 3.14 \cdot (2 \times 10^{-4} m)^2 = 13 \times 10^{-8} m^2$$

We see that the area times solid angle theorem gives us a larger area. It is more realistic than the result of the approximate paraxial theory.

8. *Point source illuminating a screen.* A point source is emitting $I_v = 10$ lm/sr or 10 candelas. How much light (in lumens) is arriving at a screen with radius $r = 1$ cm if the screen is a distance $R = 30$ cm away. Convert lumens to watts.

Solution

$$\text{Solid angle:} \quad \frac{\pi r^2}{R^2} = \frac{3.14 cm^2}{(30^2) cm^2} = 0.0035 \text{ sr}$$

$$P_v = 10 \frac{\text{lumen}}{\text{sr}} (.0035) \text{sr} = 0.035 \text{ lumen}$$

in Watts $0.035 \cdot 0.00146 = 5.1 \times 10^{-5}$ W

9. *A 40 W light bulb.* A 40 W light bulb considered as a point source has approximately 40 lm/sr, or 40 candelas. How large are the luminous flux P_v and radiant flux P into a solid angle having an area 10 cm × 10 cm and a distance of 2 m?

Solution

$$d\Omega = \frac{0.1 \cdot 0.1\ m^2}{4m^2} = \frac{0.01}{4}\ sr = 25x10^{-4} sr$$

$$P_v = 40\frac{lumen}{sr} \cdot 25x10^{-4}\ sr = 0.1\ lumen$$

$$P = 0.00146\ \frac{W}{lumen}\ 0.1\ lumen$$

$$= 0.000146\ W = 1.46x10^{-4}\ W$$

10. *A light bulb emitting different values of I_v in different directions.* We approximate a light bulb as a point source. The bulb emits 6 lm/sr in a direction perpendicular to the axis and 10 lm/sr in a direction parallel to the axis. How many lumens are incident on a screen with radius $r = 1$ cm, placed 30 cm away in both directions? Convert lumens to watts.

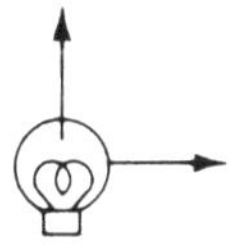

Solution

From Problem 8 we have for the 10 $\frac{lumen}{sr}$ direction

$$P_{v(10)} = 0.035\ lumen\ or\ 5.1\ x\ 10^{-5}\ W$$

For the 6 $\frac{lumen}{sr}$ direction we get

$$P_{v(6)} = 6\ \frac{lumen}{sr}\ (0.0035)\, sr = 0.021\ lumen$$
$$= 3x10^{-5}\ W$$

11. *Area illuminating a point receiver.* In most cases, the area of the receiver (detector) is known. If not, we can consider the illuminance $E_v = L_v \pi \sin^2 \alpha$. Consider a cathode-ray tube (CRT) with area 20 cm × 20 cm and luminance $L_v = 5 \times 10^4\ lm/m^2 sr$ placed a distance of 1 m from a detector with area 1 mm^2. Calculate the luminous flux P_v and the illuminance E_v at the detector.

Solution

$$P_v = 5.10^4 \frac{\text{lumen}}{m^2 sr} \quad 1x10^{-6}\ m^2\ (3.14 \cdot \sin^2 \alpha) sr$$

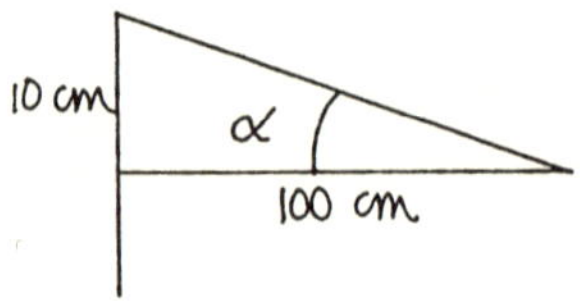

$$\tan \alpha = \frac{1}{10} \qquad \alpha = 5.7°$$

$$\sin^2 5.7 = 0.01$$

$$P_v = 5 \cdot 1 \cdot (3.14) \cdot (0.01) x10^{-6} x10^4 \frac{\text{lumen}}{m^2 sr}\ m^2\ sr$$

$$= 0.157x10^{-2}\ \text{lumen}$$

$$E_v = \frac{0.157x10^{-2}\ \text{lumen}}{1x10^{-6}\ m^2} = 0.157x10^{+4} \frac{\text{lumen}}{m^2}$$

12. *Infinite line source.* Show that for an infinite line source, the luminous flux

$$P_v = L_v\, da' \left(c \int_{-\infty}^{+\infty} \frac{db \sin\theta}{s^2} \right)$$

results in

$$P_v = L_v\, da'\, c \frac{2}{R}$$

where R is the shortest distance from da' to the line source. The solid angle is $c(2/R)$ in steradians.

Solution

We have the relations $b = R \cos\theta$, $db = -\frac{R}{\sin^2\theta}\, d\theta$

$$s \sin\theta = R$$

and

get from eq. (11.46) $P_v = -L_v\, da' \frac{c}{R} \int_{\theta_1}^{\theta_2} \sin\theta\, d\theta$

for $b = -\infty$ we have $\theta_1 = 180°$, for $b = +\infty$ $\theta_2 = 0$

that is

$$P_v = -L_v\, da' \frac{c}{R} \int_{180}^{0} \sin\theta\, d\theta$$

$$= +L_v \, da' \, \frac{c}{R} \Big| +\cos\theta \Big|_{180}^{0}$$

$$= +L_v \, da' \, \frac{c}{R} \, [+1 - (-1)] = L_v \, da' \, \frac{c}{R} \cdot 2$$

13. *Finite line source.* A line source is symmetrically placed with respect to the receiver area da' (with shortest distance R to the receiver) and has an extension $-B/2$ to $+B/2$ so that θ has the corresponding values θ_1 and θ_2. Calculate P_v. ($P_v = L_v \, da'(c/R) \cdot 2\cos\theta_2$.)

<u>Solution</u>

With the calculation of problem 12 and eq. 11.46 we have

$$P_v = -L_v \, da' \, \frac{c}{R} \int_{\theta_1}^{\theta_2} \sin\theta d\theta = -L_v \, da' \, \frac{c}{R} \Big| -\cos \Big|_{\theta_1}^{\theta_2}$$

$$P_v = L_v \, da' \, \frac{c}{R} \, (\cos\theta_2 - \cos\theta_1)$$

we have from symmetry $\theta_2 + \theta_1 = 180°$

$$\cos\theta_1 = \cos(180 - \theta_2)$$
$$= -\cos\theta_2$$

$$P_v = L_v \, da' \, (\frac{c}{R} \, 2\cos\theta_2)$$

14. *Fluorescent lamp and point receiver.* Consider a fluorescent lamp of luminance 0.5×10^4 lm/m^2 sr having a width of 3 cm and a length B of 2 m. Calculate the luminous flux P_v received by a detector with area 1 mm^2 symmetrically placed at a distance R of 1 m.

<u>Solution</u>

From problem 12

$$P_v = L_v \, da' \, (\frac{c}{R} \, 2\cos\theta)$$

$$L_v = 0.5 \times 10^4 \, \frac{\text{lumen}}{\text{sr m}^2}$$

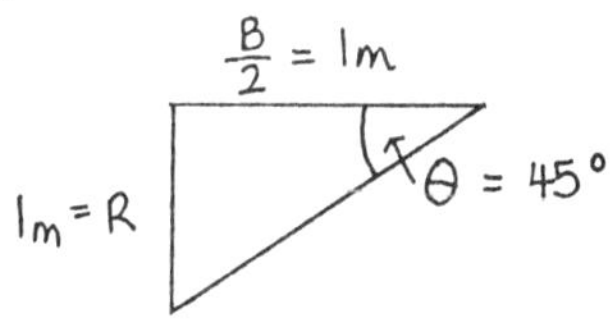

$$da' = 10^{-6} \, m^2$$

$$(\frac{c}{R} \, 2\cos\theta) = \frac{0.03m}{1m} \, 2\cos 45° = 0.06 \cdot 0.71 \text{ sr}$$
$$= 0.04 \text{ sr}$$

$$P_v = 0.5 \times 10^4 \, \frac{\text{lumen}}{\text{sr m}^2} \; 10^{-6} \; m^2 \cdot 0.04 \text{ sr}$$

$$= 0.5 \cdot 0.04 \times 10^{-2} \text{ lumen} = 2 \times 10^{-4} \text{ lumen}$$

15. *Solar panel.* A solar panel with dimensions 1 m × 2 m is illuminated by the sun. Assume that the solar detectors are sensitive only in the wavelength region where the eye is sensitive.

a. Calculate the luminous flux for the case when the sun is at its zenith position and for the later cases when the angle with the zenith position is 30° and 60°.

b. Calculate the total power arriving at the panel when the sun is at its zenith position and when the sun is 30° and 60° off.

Solution

(a) We have the area of the Sun and the area of the panel small compared to the distance to the Sun. Therefore we can use

$$P_v = L_v \frac{da\, da'}{R^2}$$

we write

$$P_v = \underset{\text{of sun}}{(L_v)} \quad \underset{\text{d}\Omega\text{ from panel to sun}}{\left(\frac{da}{R^2}\right)} \quad \underset{\text{panel}}{(da')}$$

$P_v = 2x10^9 \dfrac{\text{lumen}}{\text{sr m}^2}$, $\dfrac{da}{R^2} = 6x10^{-5}\text{sr}$

$da' = 2\ m^2$

$P_v = 2 \cdot 6 \cdot 2x10^9 x10^{-5}$ lumen

$= 24x10^4$ lumen

Multiplication with cos 30° and cos 60° respectively gives us

at 30° $P_v = 20.8x10^4$ lumen

at 60° $P_v = 12x10^4$ lumen

(b) We use now $L = 2.25x10^7 \dfrac{W}{\text{sr m}^2}$ and have

$P_v = 27x10^2 W$

at 30° $P_v = 24.4x10^2$ W

at 60° $P_v = 13.5x10^2$ W

If we would convert the figures from problem A into watt, we would get smaller values than the figures in problem B. In problem A a smaller wavelength interval is employed.

12 Image Formation and Light Throughput

1. *Power through lens and entrance and exit pupils.* A lens of radius $r' = 5$ cm and focal length $f = 20$ cm produces an image of a self-emitting object of radius $r = 1$ cm, and the lens is placed at a distance of 60 cm. Assume $L = 1$ (W/m^2 sr).

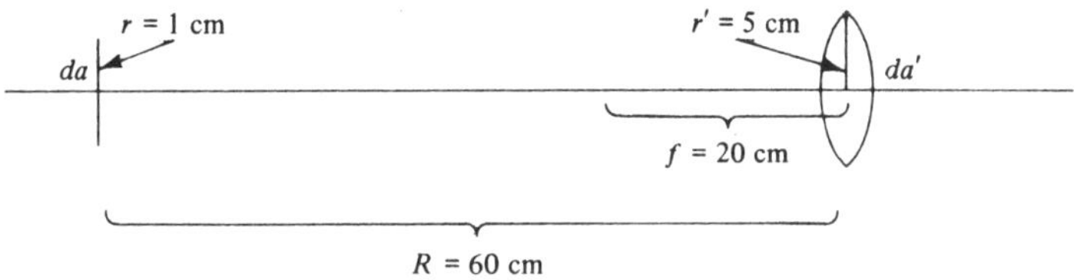

(a)

a. Calculate the power incident on the lens.
b. A round aperture of radius 0.8 cm is placed 40 cm to the left of the lens. Calculate the power P_1 passing the entrance pupil.

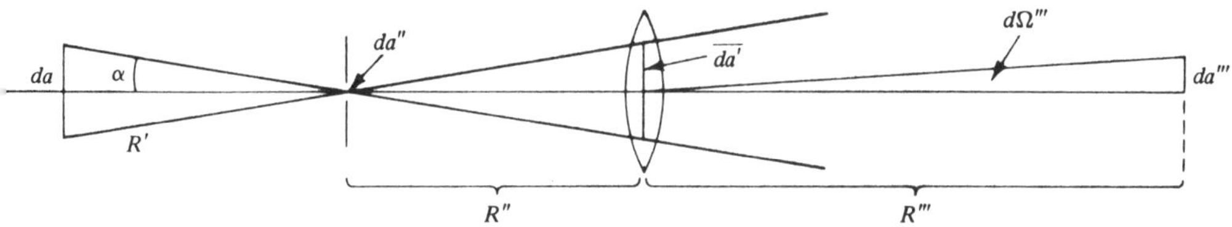

(b)

c. Express the power P_1 using da'' and the solid angle da/R'^2 and calculate the half-angle α.
d. Using the half-angle α, calculate the solid angle $d\Omega$ extending from the entrance pupil to the lens. Express the solid angle as $d\Omega = \overline{da'}/R''^2$ and calculate $\overline{da'}(R'' = 40$ cm).
e. Calculate the product of the area $\overline{da'}$ and the solid angle from the lens to the exit aperture. If multiplied by L, we must get P_1.

Solution

(a)

$$P = L\, da \frac{da'}{R^2}$$

$$da = (1\text{cm})^2 \pi = 3.14 \times 10^{-4} m^2$$

$$da' = (5\text{cm})^2 \pi = 78.5 \times 10^{-4} m^2$$

$$R^2 = (.6)^2 m^2 = 0.36 m^2$$

$$P = \frac{3.14 \cdot (78.5)}{0.36\ m^2} \cdot 10^{-4} m^2\ 10^{-4} m^2\ L$$

$$= 685 \times 10^{-8}\ Lm^2 sr = 6.85 \times 10^{-6}\ W$$

(b)

$$P_1 = L\, da \frac{da''}{R'^2}$$

$da = 3.14 \times 10^{-4} m^2$

$da'' = \pi (0.8)^2 x10^{-4} m^2 = 2x10^{-4} m^2$

$R'^2 = (0.2)^2 m^2 = 0.04 m^2$

Take L = 1

$$\frac{da''}{R'^2} = d\Omega'' = \frac{\pi \cdot 0.64x10^{-4} m^2}{0.04 m^2} = 50x10^{-4}$$

$$= L\ 158 \times 10^{-8} m^2 sr = 1.58 \times 10^{-6}\ W$$

(c)

$$P_1 = L\, da'' \frac{da}{R'^2} = 2x10^{-4} m^2 \ \frac{3.14x10^{-4} m^2}{0.04 m^2}$$

$$= 2x10^{-4} m^2 \cdot (78.5x10^{-4}) = 157x10^{-8} m^2$$

$$\tan \alpha = \frac{1cm}{20cm} \Rightarrow \alpha = 2.86°$$

(d)

$$d\Omega = \pi \sin^2 \alpha = 3.14 \cdot 0.00256$$

$$= 0.00782$$

$$= 78.2x10^{-4} sr$$

$$\overline{da'} = 78.2x10^{-4} \cdot (0.4)^2 m^2$$

$$= 12.5x10^{-4} m^2$$

(e) The area of the exit pupil is positioned at R" = R‴ equal to 2 times the focal length and we have da" = da‴

$$d\Omega''' = \frac{da'''}{R'''^2} = \frac{da''}{R''^2} = \frac{2 \times 10^{-4} m^2}{0.16 m^2}\ sr$$

$$\overline{da'}\, d\Omega''' = 12.5 \times 10^{-4}\ m^2 \cdot \frac{2 \times 10^{-4}}{0.16}\ sr$$

$$= 156.25 \times 10^{-8}\ m^2\ sr$$

$$P = L\, \overline{da'}\, d\Omega''' = 1.56x10^{-6}\ W$$

2. *Entrance and exit pupils.* Consider a self-emitting object of radius 1 cm placed 60 cm to the left of a lens of radius 5 cm and focal length 20 cm. A round aperture of radius 0.8 cm is introduced at a distance 30 cm to the left and serves as the entrance pupil. Calculate the solid angle of the light leaving the exit pupil.

Solution

(Some numbers for this problem are the same as for problem 1.)

We have $da\, d\Omega'' = da_{ex}\, d\Omega_{ex}$

We have for $da = 2 \times 10^{-4} m^2$

$$d\Omega'' = \frac{\pi \times (0.008)^2}{(.3)^2} = 22.3 \times 10^{-4} \text{ sr}$$

$$da \cdot d\Omega'' = 44 \times 10^{-8} \text{ m}^2\text{sr}$$

To calculate da_{ex} we use the thin lens equation for the position

$$\frac{1}{-x_o} + \frac{1}{x_o} = \frac{1}{f}$$

$$\frac{1}{30\text{cm}} + \frac{1}{x_i} = \frac{1}{20\text{cm}} \rightarrow \frac{1}{x_i} = \frac{1}{20} - \frac{1}{30} = \frac{30-20}{600}$$

$$\frac{1}{x_i} = \frac{1}{60} \qquad x_i = 60\text{cm}$$

the ratio of the radius of da" to the radius of da_{ex} is equal to 30:60

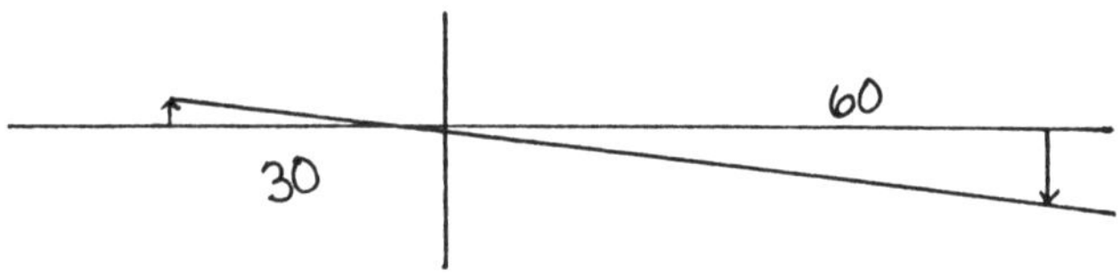

$$da_{ex} = \pi(1.6)^2\text{cm}^2 = \pi \times 2.56 \times 10^{-4}\text{m}^2 = 8.04 \times 10^{-4}\text{m}^2$$

$$d\Omega_{ex} = \frac{44 \times 10^{-8}\text{m}^2\text{sr}}{8.04 \times 10^{-4}\text{m}^2} = 5.47 \times 10^{-4}\text{sr}$$

that is 1/4 of dΩ"

3. *Field lens and condenser lens.* Consider two lenses of radius 5 cm and focal length 20 cm. An object of radius 1 cm is placed 40 cm to the left of the first lens. The distance between the lenses is 80 cm. The image after the second lens is 40 cm to the right of it.

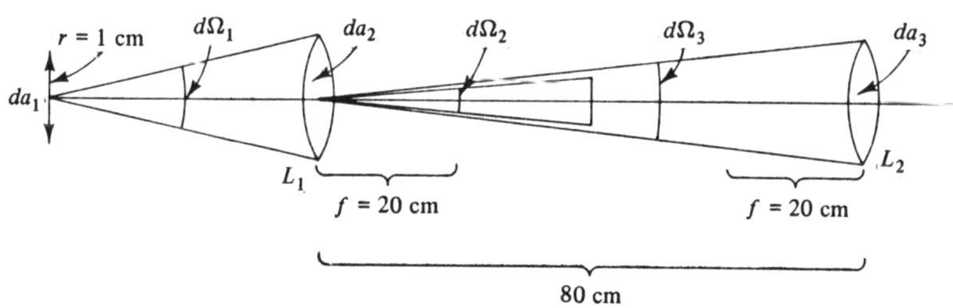

a. Calculate the product $da_1\, d\Omega_1$ for the power incident onto the first lens.
b. Calculate the solid angle into which the light travels from lens L_1.
c. How large is the solid angle $d\Omega_3$ from lens L_1 to lens L_2? Does all the light from the object arrive at lens L_2?

d. The object now has a 5 cm radius. How large is the solid angle into which the light travels from lens L_1? How much light does not arrive at lens L_2?
e. A field lens is placed at the image position. Show that all the light arrives at L_2.

Solution

(a) $da_1 = 3.14\text{x}10^{-4}\text{m}^2$ (from problem 1)

We use the small angle approximation $d\Omega = \frac{\text{area}}{(\text{distance})^2}$

$$d\Omega_1 = \frac{da_2}{(40\text{cm})^2} = \frac{78.5\text{x}10^{-4}\text{m}^2}{0.16\text{m}^2} = 490\text{x}10^{-4}\text{sr}$$

$$= 4.9\text{x}10^{-2}\text{sr}$$

$$da_1 \cdot d\Omega_1 = 15.4\text{x}10^{-6}\text{sr m}^2$$

(b) $da_2\ d\Omega_2 = 15.4\text{x}10^{-6}\text{sr m}^2$

$da_2 = 78.5\text{x}10^{-4}\text{m}^2$

$$d\Omega_2 = \frac{15.4\text{x}10^{-6}\text{sr m}^2}{78.5\text{x}10^{-4}\text{m}^2} = .196\text{x}10^{-2}\text{sr}$$

(c) $da_3 = da_2 = 78.5\text{x}10^{-4}\text{m}^2$

$$d\Omega_3 = \frac{78.5\text{x}10^{-4}\text{m}^2}{(80\text{cm})^2} = \frac{78.5\text{x}10^{-4}}{0.64} = 122\text{x}10^{-4}\text{sr}$$

$$= 1.22\text{x}10^{-2}\text{sr}$$

all the light is entering L_2.

(d) $da_1' = 78.5\text{x}10^{-4}\text{m}^2$

$d\Omega_1' = d\Omega_1 = 4.9\text{x}10^{-2}\text{sr}$

$da_1' \cdot d\Omega_1 = 384.7\text{x}10^{-6} = 3.85\text{x}10^{-4}\text{m}^2\text{sr}$

$d\Omega_2 = 4.9\text{x}10^{-2}\text{sr}$

compared to $d\Omega_3 = 1.22\text{x}10^{-2}\text{sr}$

fraction $\frac{4.9}{1.22} = 4$ 3/4 of the light

Solution (cont'd)

(e)

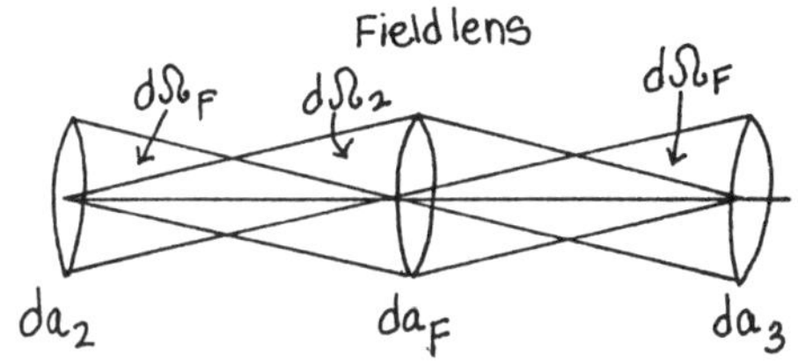

We have $da_2\ d\Omega_2 = da_F\ d\Omega_F = da_3\ d\Omega_F$

4. *Numerical aperture of the microscope.* In Chapter 9, we found that the smallest distance the microscope can resolve is

$$I = \frac{\lambda}{2n \sin \alpha}$$

where α is the half-angle of the solid angle of the light from the object to the objective lens. The solid angle enters the expression through the usefully illuminated diameter of the objective lens. The index n is the refractive index of the immersion oil. Compare the resolvable distance for the case where there is no condenser ($\alpha = 10^\circ$) and the case where there is a condenser and immersion oil ($\alpha = 70^\circ$ and $n = 1.5$).

Solution

$$y_{\text{no cond.}} = \frac{\lambda}{2 \sin 10^\circ} = \frac{\lambda}{0.35}$$

$$y_{\text{cond.}} = \frac{\lambda}{2 \cdot 1.5 \cdot \sin 70^\circ} = \frac{\lambda}{2.8}$$

We see that for these drastic different values the resolution is improved by a factor of about 10.

5. Consider a telescopic system with an objective lens of diameter $d = 5$ cm and an ocular of $d' = 1$ cm.
   a. Show that

$$\frac{d}{d'} = \left|\frac{f_1}{f_2}\right| = \frac{\beta}{\alpha}$$

   and give the angular magnification β/α, where f_1 is the focal length of the objective, f_2 that of the ocular, α is the angular size seen on the retina without the telescope, and β the angular size seen by the ocular of the first image (produced by the objective lens).

b. If $f_1 = 10$ cm and $f_2 = 2$ cm, then we satisfy the above conditions. What happens if $f_1 = 20$ cm and $f_2 = 2$ cm? Give d/d''. Since d corresponds to the entrance pupil, what happens to the exit beam and the exit solid angle?

12.5 Solution

(a) From chapter 1 we have $\frac{\beta}{\alpha} = \left|\frac{f_1}{f_2}\right|$ and from the area times solid angle consideration $\frac{\beta}{\alpha} = \frac{d}{d'}$,

It follows $\frac{d}{d'} = \left|\frac{f_1}{f_2}\right|$

$\frac{d}{d'} = 5 = \left|\frac{f_1}{f_2}\right|$ and we have $\frac{\beta}{\alpha} = 5$

(b) $\frac{d}{d''} = 10$, since d is the diameter of the entrance pupil, d" is the diameter of the exit pupil, that is d" = .5cm and that is at the same time the diameter of the exit beam. The area of the ocular is not fully used. The solid angle of the exit beam is enlarged by the factor $\frac{(d')^2}{(d'')^2} = \frac{1}{.25} = 4$

6. *Orientation of the image of a lens.* Consider the real image of a real object of a positive lens.

a. Give the transformation in coordinates from object to image if a coordinate system is used as shown in the following figure.

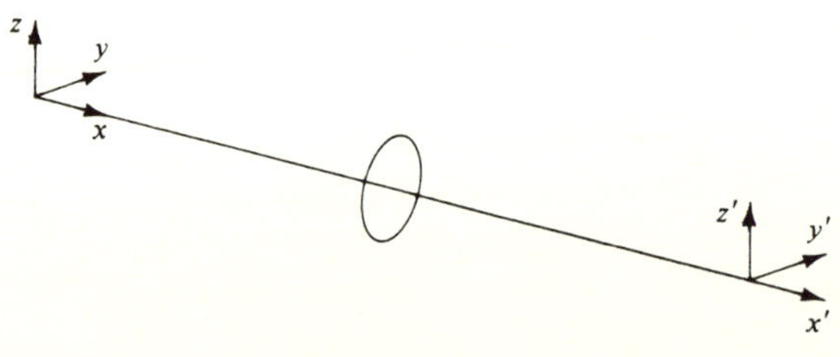

(a)

b. If we have an object of the type what is the configuration of the image?

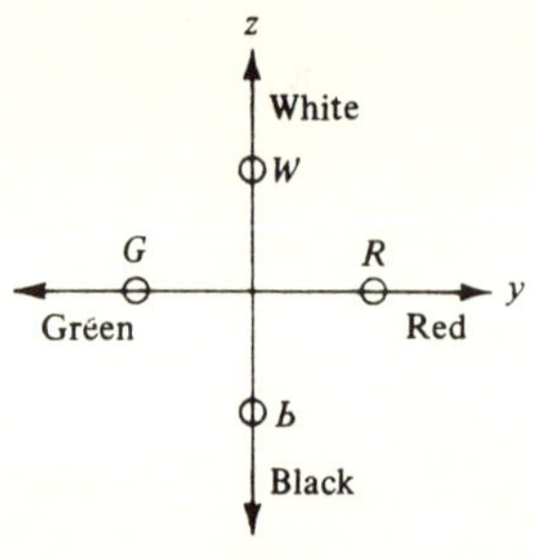

(b)

c. Can we obtain the image configuration from the object configuration by rotation of the object?
d. The object contains a right-handed rotation configuration B-R-W-G. Does the image also have a right-handed configuration?

Solution

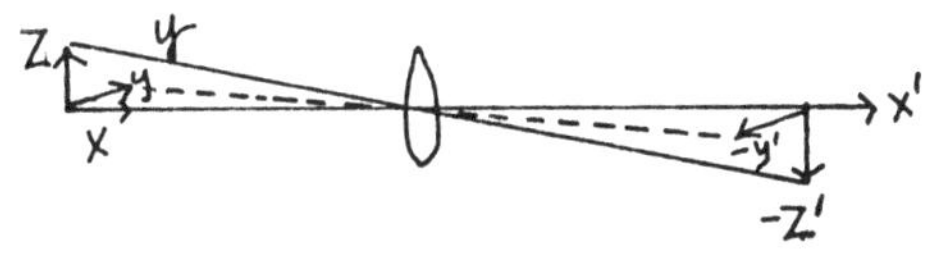

(a) $x \to x'$, $y \to -y'$, $z \to -z'$

(b)

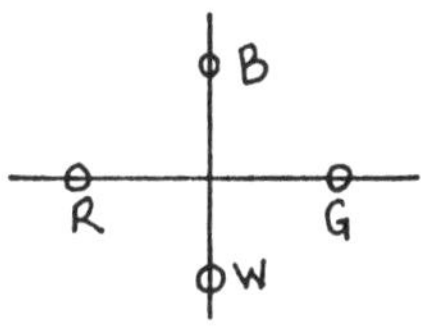

(c) Rotation by 180° of the object configuration results in the image configuration.

(d) Image W-G-B-R → B-R-W-G

It is the same, right handed.

7. *Orientation of the image produced by a mirror*
   a. Give the transformation of the coordinates from a real object and the virtual image produced by a mirror. If we trace back the image to the object with parallel rays, reflected at the mirror, we find the same transformation. (Take the angle of incidence at 45° to the normal.) make a sketch.
   b. If x, y, z is a right-handed coordinate system, (i) what is the coordinate system at the virtual image and (ii) what is the coordinate system of the reflection projection (right- or left-handed?)
   c. If we place the object

$$\begin{array}{ccc} & \uparrow z & \\ & W & \\ G & & R \rightarrow y \\ & B & \end{array}$$

   in the right-handed coordinate system of the y, z plane, how does the image configuration appear when we look into the light from the source?
   d. Can we rotate the object configuration to get the image configuration of (c)?

Solution

(a) x → x', y → y', z → z'

a) 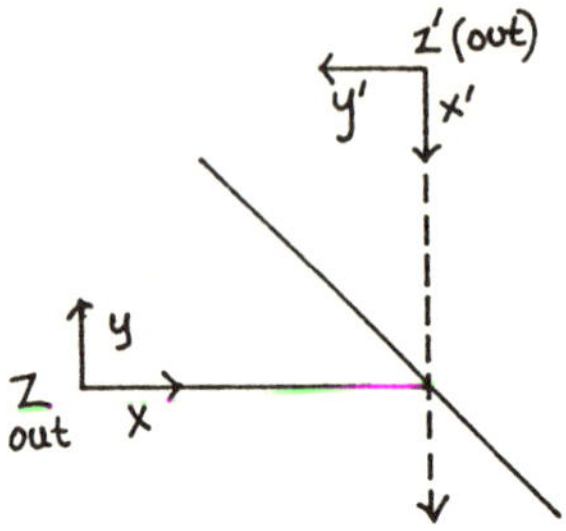

b) 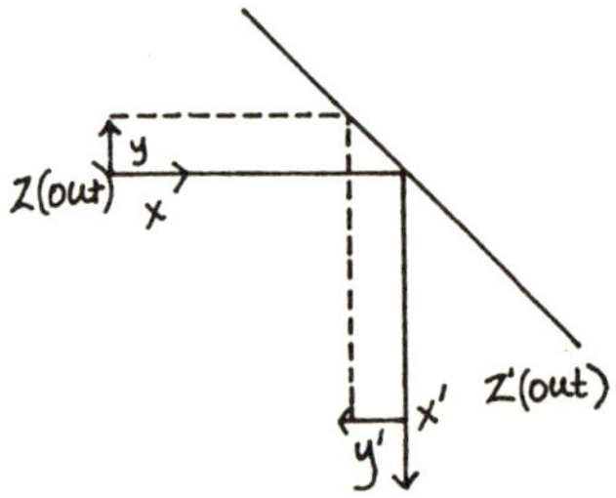

(b) i x - y - z right handed

x'- y'- z' left handed

ii x - y - z right handed

x'- y'- z' left handed

(c)

	z				z'	
	↑				↑	
	W				W	
G		R → y	[Mirror]	y' ← R		G
	B				B	

(d) No.

8. *One Porro prism, vertical extension*

a. Consider the Porro prism shown in the figure, where two reflecting surfaces make a 90° angle. If we look against the travel direction of the light, the object has the configuration

W

G *R*

B

Give the image configuration looking against the travel direction of the light and looking with the light. (Left and right in reference to: perpendicular to the plane of incidence.)

b. Which of the two cases in (a) can be obtained by rotating the object configuration by 180°?

c. How is the direction of the light changed with respect to the vertical plane (plane of incidence)?

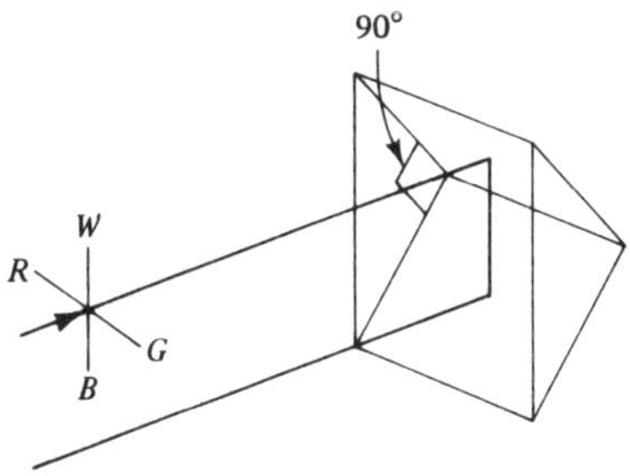

<u>Solution</u>

(a)

```
       W
   G       R
       B
object looked "against"
```

```
       B
   R       G
       W
image looked "against"

       B
   G       R
       W
image looked "with"
```

(b) If we look against the light.

(c) The direction of the light is changed by 180° in the vertical plane.

9. *One Porro prism, horizontal extension*
   a. Consider the Porro prism shown in the figure, where two reflecting surfaces make an angle of 90°. If we look against the light, the object has the configuration

$$\begin{array}{ccc} & W & \\ G & & R \\ & B & \end{array}$$

   Give the image configuration looking against and with the direction of the travel of the light. (Left and right in the plane of incidence when W is up and B is down.)
   b. What is the rotation of the object configuration to obtain the configuration "with" and "against" the direction of the light?
   c. What does the prism do to the light if it is not changing the configuration of the object? [Case: against the direction of the light; see part (b).]

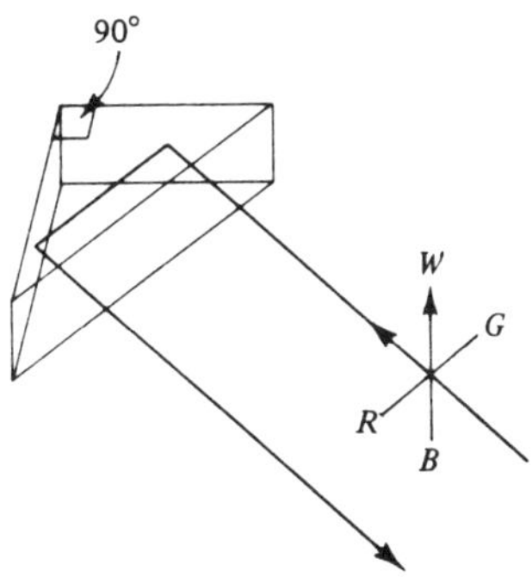

Solution

(a)

W G R B	against →	G	W B	R
	with →	R	W B	G

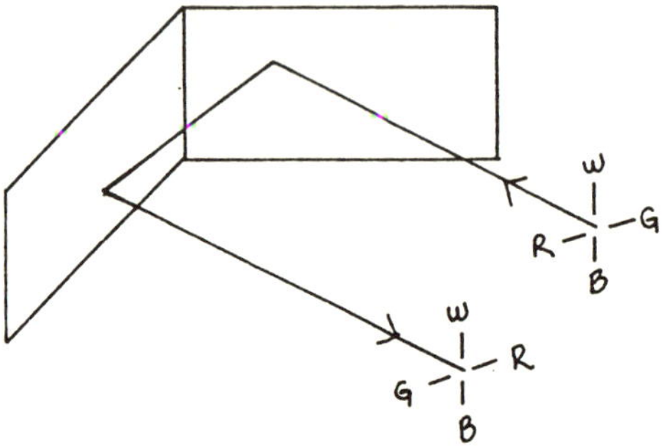

(b) Against : 0 rotation

With: cannot be obtained by rotation

(c) It changes the direction of the emerging light by 180° with respect to the incident light in the horizontal plane. (Plane of incidence)

10. *The double Porro prism, one vertical, one horizontal*
   a. Use the two Porro prisms and give the configuration of the object after one prism; after two prisms. Always look against the direction of the light.

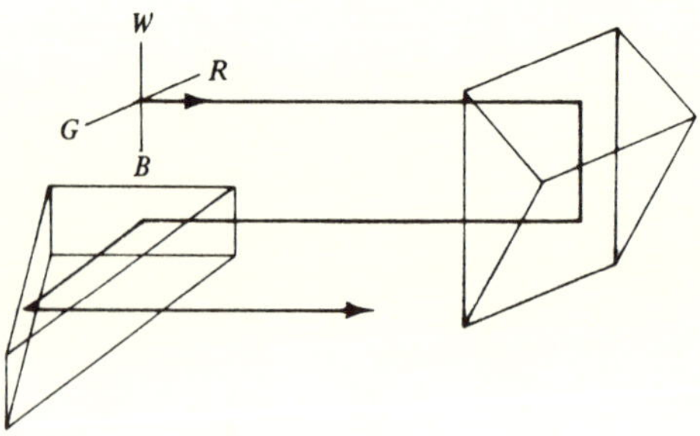

   b. Can we obtain the image configuration from the object configuration by rotation?
   c. What does the first prism do to the direction of the light compared to what the second prism does?
   d. Compare the action of the double Porro prism to the action of the positive lens, with respect to the imaging of real object to real image.

Solution

(a)	Object	W G R B
	After one	B R G W
	After two	B R G W

(b) Rotation by 180°.

(c) The first changes the direction of the light by 180°. The second changes the direction of the light again by 180°

(d) Both prisms together do the same job as the lens does and the exit light travels parallel to the incident light.

11. *Configuration changes and group theory.* We can make a list of the action of a lens, a mirror, and various Porro prisms. Looking always into the direction of the light, we have the following:

Object Configuration	Operator	Image Configuration
W *G* *R* *B*	*L* (lens)	*B* *R* *G* *W*
W *G* *R* *B*	*M* (mirror)	*W* *R* *G* *B*
W *G* *R* *B*	*PV* (Porro vertical)	*B* *R* *G* *W*
W *G* *R* *B*	*PH* (Porro horizontal)	*W* *G* *R* *B*
W *G* *R* *B*	*DP* (double Porro)	*B* *R* *G* *W*

There is no interchange of *B* or *W* with *R* or *G*. Therefore, we can characterize the operations as follows:

$$L: (B \leftrightarrow W)(R \leftrightarrow G)$$

$$M: (R \leftrightarrow G)$$

$$PV: (B \leftrightarrow W)(R \leftrightarrow G)$$

$$PH: \text{No change}$$

$$DP: (B \leftrightarrow W)(R \leftrightarrow G)$$

We have three different operations:

$$L: (B \leftrightarrow W)(R \leftrightarrow G)$$

$$M: (R \leftrightarrow G)$$

$$PHE: (E) \text{ (no action)}$$

If we use these operations one after another (a process called multiplication), we find that all products, regardless of how often we multiply the operations with one another, are

$$L: (B \leftrightarrow W)(R \leftrightarrow G)$$

$$M: (R \leftrightarrow G)$$

$$K = L \cdot M: (B \leftrightarrow W)$$

$$E: (E) \text{ (no action)}$$

The operation K is the product of L and M, because twice the operation $(R \leftrightarrow G)$ results in "doing nothing," equivalent to operation (E), and E times E is E.

a. Fill in the multiplication table, where the elements across the upper row come first, the elements in the left column come second.

b. The elements E, K, L, and M form a finite group (called a Klein four-group). To form a group, we must have
   (1) A multiplication rule: $A \cdot B = C$ for all elements in the group
   (2) A distribution law: $(A \cdot B)C = A(B \cdot C)$
   (3) A unit element and an inverse element
   If $A \cdot B = B \cdot A$, the group is called an Abelian group. Show that all this is fulfilled by the Klein four-group. (Note: It was not necessary to say which of the elements should be multiplied first.)

c. A subgroup contains fewer elements than the full group. How many subgroups does the Klein four group have?

Solution

(a)

	E	K	L	M
E	E	K	L	M
K	K	E	M	L
L	L	M	E	K
M	M	L	K	E

We see that the multiplication table is symmetrically arranged with respect to the diagonal. The product of two elements is always the third, except if E is involved.

(b) 1) Multiplication see table.

2) By inspection $(KL)M = K(LM) = M \cdot M = K \cdot K = E$ for all such products, if E is involved it reduces to the multiplication table.

3) Unit element E is there

$K \cdot K = L \cdot L = M \cdot M = E$

Inverse element: Each element is its own inverse since we have e.g. $K \cdot K = E$.

(c) The subgroups are E, K

E, L

E, M

E itself is a trivial subgroup, counting it also, we have four subgroups.

12. *Homomorphism of two groups.* With respect to the property "left-handed coordinate system" or "right-handed coordinate system," we have the following properties of the operations:

$$L\text{: No change } (e)$$
$$M\text{: } l \leftrightarrow r \ (a)$$
$$K\text{: } l \leftrightarrow r \ (a)$$
$$E\text{: No change } (e)$$

We have the relation that L and E act as a unit operation (e) with respect to direction of coordinate system changes, and M and K act as an operation of change denoted by (a).

Show that (e), (a) form a group. The group (e, a) has two elements; its order is 2. The Klein four group has four elements; its order is 4. The relation between the Klein four-group and the two-group (e, a) is a process of mapping. We have mapped the four-group into the two-group, and more than one element of the four-group corresponds to one element of the two group. Properties of group theory are useful to bring order into configurations and to answer the question, How many different image configurations can be obtained with a certain number of optical (operator) elements?

Solution

Multiplication table.

	e	a
e	e	a
a	a	e

1) Multiplication O.K.

2) Distributive laws are O.K.

3) Unit element, inverse element O.K.

13. *The Dove prism.* We will replace the light pass in the Dove prism by three reflections; see figure (B is below the page, W above; the mirror is perpendicular to the page).

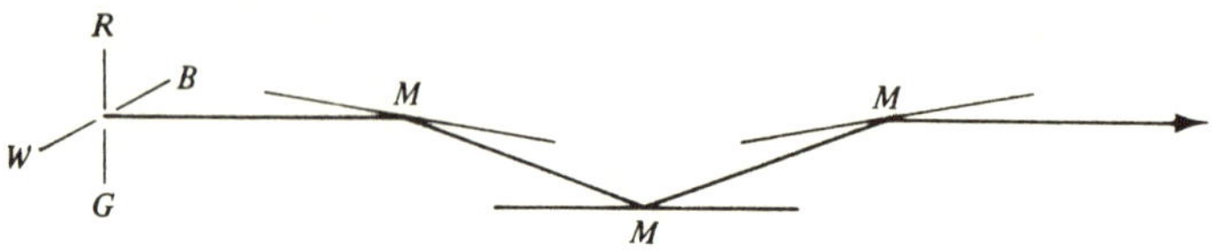

We place an object

W

$G \qquad R$

B

to the left. The configuration is for looking against the direction of travel of the light.

a. Using what we have learned about group theory, show that the effect of three reflections is equivalent to one reflection.

b. Give the image configuration.

c. We now turn the prism by 90° into the following position:

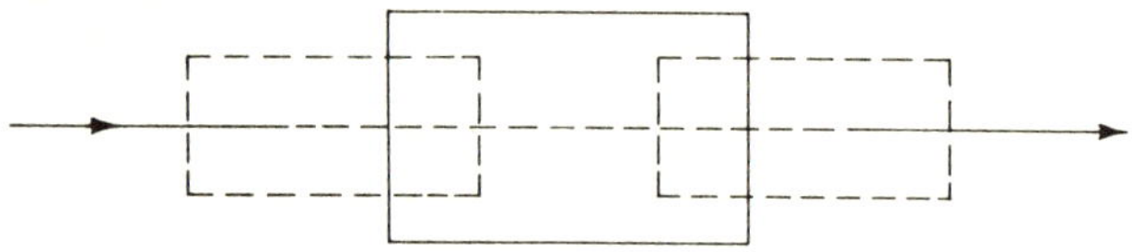

Give the image formation.

d. How does the image look if we look at it with the light (not against the light)?

e. Compare the result from part (d) with the object looked at against the light, and indicate the rotation of 180° by the object to get this image.

<u>Solution</u>

(a) The reflection M forms a subgroup with the unit element of the full Four Group. Multiplication of M • M • M = M.

(b)

```
      object          image
        W               W
    G       R       R       G     against the light
        B               B
```

(c)

```
        W               B         after 90° rotation of
    G       R       G       R
        B               W         mirrors in direction against
                                  the light
```

(d)

W
G R
B

B
R G
W

after 90° rotation of mirrors in direction with the light

(e)

W
G R
B

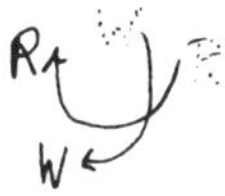

and so on gives

B
R G
W

For the special case that the prism is rotated by 90° we have seen that the image (looked at in direction with respect to the light) is rotated by 180°

13 Aberration

1. *Spherical aberration of a single refracting surface.* Consider the rays from P_1 to P_2 passing through the refracting surface at distance ρ (see following figure). In Chapter 1,

$$\frac{|s_1| + |r|}{|s_2| - |r|} = n\frac{|\zeta_1|}{|\zeta_2|}$$

(equivalent to equation with $n_1 = 1$, $n_2 = n$)

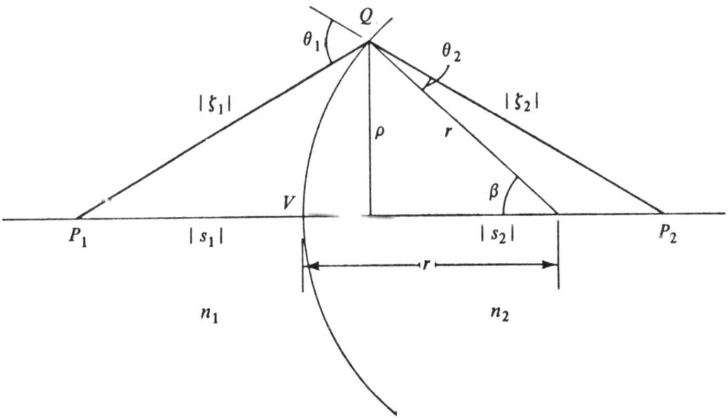

Now we do not set $s_1 \simeq \zeta_1$ and $s_2 \simeq \zeta_2$, but introduce in the following a better approximation. We neglect the absolute bars. Using the law of cosines, we get

$$\zeta_1^2 = r^2 + (s_1 + r)^2 - 2r(s_1 + r)\cos\beta$$

Using

$$\cos\beta = 1 - \frac{\beta^2}{2}$$

and setting $\beta = \rho/r$ for small β, we get

$$\zeta_1^2 = r^2 + (s_1 + r)^2 - 2r(s_1 + r)\left(1 - \frac{\rho^2}{2r^2}\right)$$

a. Show that we obtain

$$\zeta_1^2 = s_1^2 + (s_1 + r)\frac{\rho^2}{r}$$

and

$$\zeta_2^2 = s_2^2 - (s_1 - r)\frac{\rho^2}{r}$$

(*Note:* For $\rho \to 0$, we are back to the approximation used in Chapter 1.)

b. Expand the root

$$\zeta_1 = \sqrt{s_1^2 + (s_1 + r)\frac{\rho^2}{r}}$$

according to

$$y = (1 + x^2)^{1/2} = 1 + \frac{x^2}{2}$$

and get

$$\zeta_1 = s_1 + \left(\frac{1}{s_2} + \frac{1}{r}\right)\frac{\rho^2}{2}$$

$$\zeta_2 = s_2 + \left(\frac{1}{s_2} - \frac{1}{r}\right)\frac{\rho^2}{2}$$

We introduce ζ_1 and ζ_2 to the first equation.

c. Show that

$$\frac{s_1 + r}{s_2 - r} = n\frac{s_1 + [(1/s_1) + (1/r)](\rho^2/2)}{s_2 + [(1/s_2) - (1/r)](\rho^2/2)}$$

can be written as

$$\frac{1}{s_1} + \frac{n}{s_2} + \frac{1 - n}{r} = \left(\frac{1}{r} + \frac{1}{s_1}\right)\left(\frac{1}{r} - \frac{1}{s_2}\right)\left(\frac{n}{s_1} + \frac{1}{s_2}\right)\frac{\rho^2}{2}$$

d. To get to the usual expression for the image formation of a single refracting surface, introduce $s_1 \to -x_o$ and $s_2 \to x_i$ and get

$$-\frac{1}{x_0} + \frac{n}{x_i} = \frac{n - 1}{r} + \left(\frac{1}{r} - \frac{1}{x_0}\right)\left(\frac{1}{r} - \frac{1}{x_i}\right)\left(\frac{1}{x_i} - \frac{n}{x_0}\right)\frac{\rho^2}{2}$$

Note that for $\rho \to 0$, we are back to our original approximation. The coefficient of $\rho^2/2$ depends on x_o and x_i. We would like to have it depend only on x_o. Then we could judge how, for the same x_o, x_i varies depending on ρ. We introduce into this coefficient our original (zero-order) approximation:

$$\frac{1}{x_i} = \frac{n - 1}{nr} + \frac{1}{nx_o}$$

e. Show that we obtain the following as our first-order approximation:

$$-\frac{1}{x_o}+\frac{n}{x_i}=\frac{n-1}{r}+\frac{n-1}{n^2}\left(\frac{1}{r}-\frac{1}{x_o}\right)^2\left(\frac{1}{r}-\frac{n+1}{x_o}\right)\frac{\rho^2}{2}$$

f. Give the position x_i for the marginal ray and the paraxial ray if $x_o \to \infty$. Use $n = 1.5$.

g. Show in a sketch the location of x_i (first order) with respect to x_i (zero order) for:

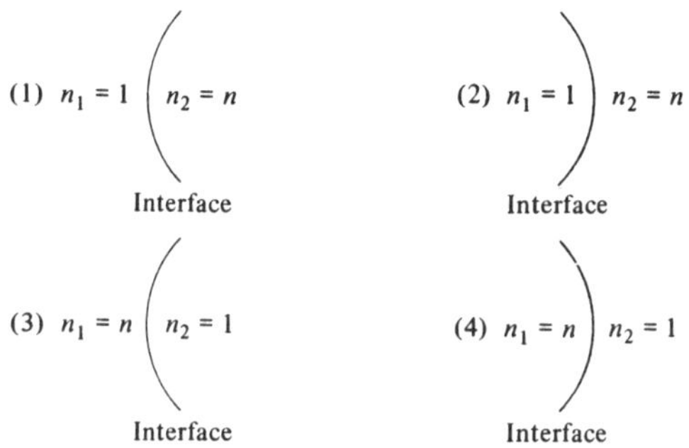

Solution

(a)

Using the Law of Cosines we get:

$$\xi_1^2 = r^2 + (s_1 + r)^2 - 2r(s_1 + r)\cos\beta$$

Letting $\cos\beta = 1 - \frac{\beta^2}{2}$ and $\beta = \frac{\rho}{r}$

$$\xi_1^2 = r^2 + (s_1 + r)^2 - 2r(s_1 + r)\left(1 - \frac{\rho^2}{2r^2}\right)$$

$$\xi_1^2 = s_1^2 + (s_1 + r)\frac{\rho^2}{r}$$

Similarly:

$$\xi_2^2 = r^2 + (s_2 - r)^2 - 2r(s_2 - r)\cos(180° - \beta)$$

$$= r^2 + (s_2 - r)^2 + 2r(s_2 - r)\cos\beta$$

We get:

$$\xi_2^2 = s_2^2 - (s_1 - r)\frac{\rho^2}{r}$$

Note that for $\rho \to 0$ we have $\begin{matrix} \xi_1 \to s_1 \\ \xi_2 \to s_2 \end{matrix}$

(b)

$$\xi_1 = \sqrt{s_1^2 + (s_1 + r)\frac{\rho^2}{r}} \qquad = s_1\sqrt{1 + (s_1 + r)\frac{\rho^2}{s_1^2 r}}$$

$$= s_1\left(1 + \frac{(s_1 + r)\rho^2}{2s_1^2 r}\right) = s_1 + \frac{(s_1 + r)\rho^2}{2s_1 r}$$

$$\xi_1 = s_1 + \left(\frac{1}{s_1} + \frac{1}{r}\right)\frac{\rho^2}{2}$$

$$\xi_2 = s_2 + \left(\frac{1}{s_2} - \frac{1}{r}\right)\frac{\rho^2}{2}$$

(c)

$$\frac{s_1 + r}{s_2 - r} = n\,\frac{s_1 + \left(\frac{1}{s_1} + \frac{1}{r}\right)\frac{\rho^2}{2}}{s_2 + \left(\frac{1}{s_2} - \frac{1}{r}\right)\frac{\rho^2}{2}}$$

$$(s_1 + r)s_2\left(1 + \left(\frac{1}{s_2} - \frac{1}{r}\right)\frac{\rho^2}{2s_2}\right) = n(s_2 - r)s_1\left(1 + \left(\frac{1}{s_1} + \frac{1}{r}\right)\frac{\rho^2}{2s_1}\right)$$

$$\left(1 + \frac{r}{s_1}\right)\left(1 + \left(\frac{1}{s_2} - \frac{1}{r}\right)\frac{\rho^2}{2s_2}\right) = n\left(1 - \frac{r}{s_2}\right)\left(1 + \left(\frac{1}{s_1} + \frac{1}{r}\right)\frac{\rho^2}{2s_1}\right)$$

$$\left(\frac{1}{r} + \frac{1}{s_1}\right)\left[\frac{1}{r} + \left(\frac{1}{s_2} - \frac{1}{r}\right)\frac{\rho^2}{2rs_2}\right] = n\left(\frac{1}{r} - \frac{1}{s_2}\right)\left[\frac{1}{r} + \left(\frac{1}{s_1} + \frac{1}{r}\right)\frac{\rho^2}{2rs_1}\right]$$

$$\left(\frac{1}{r} + \frac{1}{s_1}\right)\frac{1}{r} - n\left(\frac{1}{r} - \frac{1}{s_2}\right)\frac{1}{r} = n\left(\frac{1}{r} - \frac{1}{s_2}\right)\left(\frac{1}{s_1} + \frac{1}{r}\right)\frac{\rho^2}{2rs_1} - \left(\frac{1}{r} + \frac{1}{s_1}\right)\left(\frac{1}{s_2} - \frac{1}{r}\right)\frac{\rho^2}{2rs_2}$$

$$\left(\frac{1}{s_1} + \frac{n}{s_2} + \frac{1-n}{r}\right) = \left[n\left(\frac{1}{r} - \frac{1}{s_2}\right)\left(\frac{1}{s_1} + \frac{1}{r}\right)\frac{1}{s_1} - \left(\frac{1}{r} + \frac{1}{s_1}\right)\left(\frac{1}{s_2} - \frac{1}{r}\right)\frac{1}{s_2}\right]\frac{\rho^2}{2}$$

$$= \left(\frac{1}{r} + \frac{1}{s_1}\right)\left(\frac{1}{r} - \frac{1}{s_2}\right)\left(\frac{n}{s_1} + \frac{1}{s_2}\right)\frac{\rho^2}{2}$$

(d)

If we let $s_1 \to -x_o$ and $s_2 \to x_i$

$$-\frac{1}{x_o}+\frac{n}{x_i}=\frac{n-1}{r}+\left(\frac{1}{r}-\frac{1}{x_o}\right)\left(\frac{1}{r}-\frac{1}{x_i}\right)\left(\frac{1}{x_i}-\frac{n}{x_o}\right)\frac{\rho^2}{2}$$

(e)

By the first order approximation we have:

$$\frac{1}{x_i}=\frac{n-1}{nr}+\frac{1}{nx_o}$$

Substituting,

$$-\frac{1}{x_o}+\frac{n}{x_i}=\frac{n-1}{r}+\left(\frac{1}{r}-\frac{1}{x_o}\right)\underbrace{\left(\frac{1}{r}-\frac{n-1}{nr}-\frac{1}{nx_o}\right)\left(\frac{n-1}{nr}+\frac{1}{nx_o}-\frac{n}{x_o}\right)}_{\frac{1}{n^2}\left(\frac{n}{r}-\frac{n-1}{r}-\frac{1}{x_o}\right)\left(\frac{n-1}{r}+\frac{1}{x_o}-\frac{n^2}{x_o}\right)}\frac{\rho^2}{2}$$

$$=\frac{1}{n^2}\left(\frac{1}{r}-\frac{1}{x_o}\right)^2\left(\frac{n-1}{r}+\frac{(1-n)(1+n)}{x_o}\right)$$

$$=\frac{n-1}{n^2}\left(\frac{1}{r}-\frac{1}{x_o}\right)^2\left(\frac{1}{r}-\frac{n+1}{x_o}\right)$$

Substituting to the original Eqn. we have:

$$-\frac{1}{x_o}+\frac{n}{x_i}=\frac{n-1}{r}+\frac{n-1}{n^2}\left(\frac{1}{r}-\frac{1}{x_o}\right)^2\left(\frac{1}{r}-\frac{n+1}{x_o}\right)\frac{\rho^2}{2}$$

(f)

When the object is at infinity;

$$\frac{1}{x_i}=\frac{n-1}{nr}+\frac{(n-1)}{2n^3r^3}\rho^2 \qquad \text{for the marginal ray}$$

and

$$\frac{1}{x_i}=\frac{n-1}{nr} \qquad \text{for paraxial ray.}$$

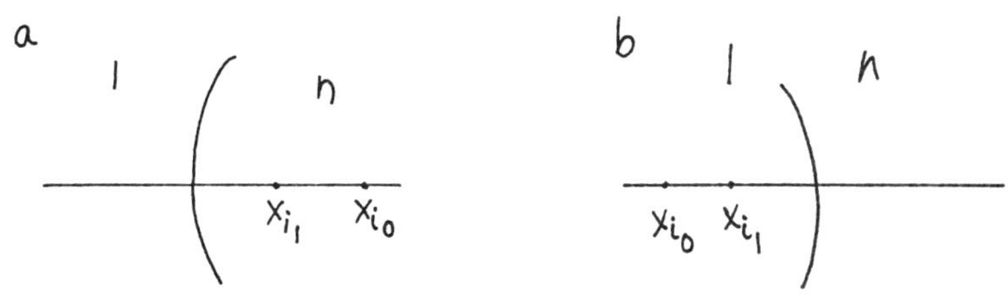

c

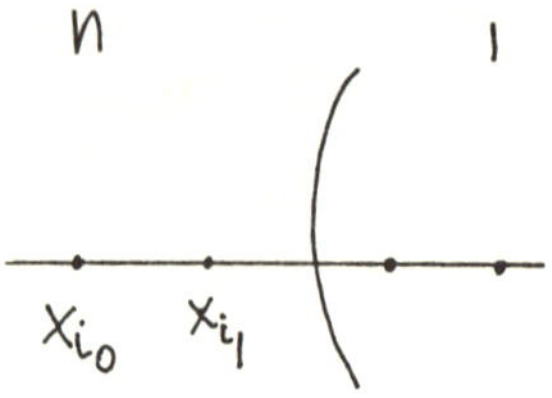

d

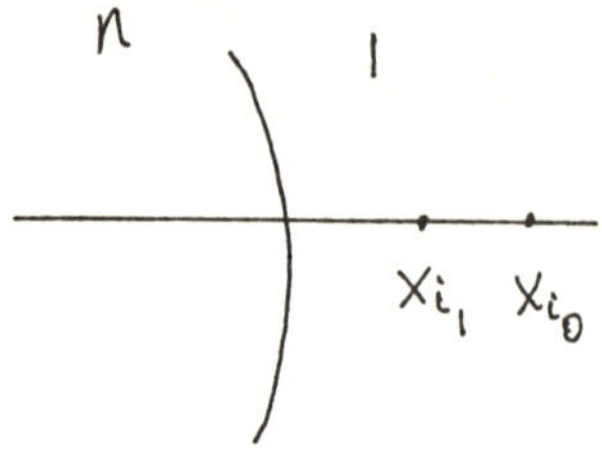

2. *Spherical aberration of a thin lens.* Let us consider the results for one interface (problem 1) where we had the refractive index $n = 1$ on the left and the refractive index n on the right. We rewrite the equations for the case where the index n is on the left of the interface and the index 1 is on the right. This means that the light is now going from the medium of index n to the medium of index 1. The result can be obtained by substituting $x_o \to x_i'$ and $x_i \to x_o'$. Calling the radius of curvature of the second surface r_2, we have

$$-\frac{1}{x_o} + \frac{n}{x_i} = \frac{n-1}{r_1} + \frac{n-1}{n^2}\left(\frac{1}{r_1} - \frac{1}{x_o}\right)^2\left(\frac{1}{r_1} - \frac{(n+1)}{x_o}\right)\frac{\rho^2}{2}$$

and

$$-\frac{n}{x_o'} + \frac{1}{x_i'} = -\frac{n-1}{r_2} - \frac{n-1}{n^2}\left(\frac{1}{r_2} - \frac{1}{x_i'}\right)^2\left(\frac{1}{r_2} - \frac{n+1}{x_i'}\right)\frac{\rho^2}{2}$$

a. Subtract these two equations.
b. Substitute the zero order into the first-order terms to eliminate $1/x_i'$:

$$\frac{1}{x_i'} = \frac{1}{x_o} + (n-1)\left(\frac{1}{r_1} - \frac{1}{r_2}\right)$$

c. Finally, obtain

$$-\frac{1}{x_o} + \frac{1}{x_i'} = (n-1)\left(\frac{1}{r_1} - \frac{1}{r_2}\right) + \frac{n-1}{n^2}\left[\left(\frac{1}{r_1} - \frac{1}{x_o}\right)^2\left(\frac{1}{r_1} - \frac{n+1}{x_o}\right)\right.$$
$$\left. - \left(\frac{1}{x_o} + \frac{n-1}{r_1} - \frac{n}{r_2}\right)^2\left(\frac{n^2}{r_2} - \frac{n+1}{x_o} - \frac{n^2-1}{r_1}\right)\right]\frac{\rho^2}{2}$$

Note: Longitudinal spherical aberration (L.S.A.) is defined as the distance differential for $\rho \neq 0$ and $\rho = 0$; that is,

$$\text{L.S.A.} = x_i(\rho \neq 0) - x_i'(\rho = 0)$$

and may be expressed as:

$$\text{L.S.A.} = \left(\frac{1}{x_i'(\rho = 0)} - \frac{1}{x_i(\rho \neq 0)}\right)x_i(\rho \neq 0)x_i'(\rho = 0)$$

Solution

(a)

$$-\frac{1}{x_o} + \frac{n}{x_i} = \frac{n-1}{r_1} + \frac{n-1}{n^2}\left(\frac{1}{r_1} - \frac{1}{x_o}\right)^2\left(\frac{1}{r_1} - \frac{n+1}{x_o}\right)\frac{\rho^2}{2}$$

$$-\frac{n}{x_o'} + \frac{1}{x_i'} = -\frac{n-1}{r_2} - \frac{n-1}{n^2}\left(\frac{1}{r_2} - \frac{1}{x_i'}\right)^2\left(\frac{1}{r_2} - \frac{n+1}{x_i'}\right)\frac{\rho^2}{2}$$

(b) $$-\frac{1}{x_o}+\frac{1}{x'_i}=(n-1)\left(\frac{1}{r_1}-\frac{1}{r_2}\right)$$

$$+\left(\frac{n-1}{n^2}\right)\left[\left(\frac{1}{r_1}-\frac{1}{x_o}\right)^2\left(\frac{1}{r_1}-\frac{n+1}{x_o}\right)-\left(\frac{1}{r_2}-\frac{1}{x'_i}\right)^2\left(\frac{1}{r_2}-\frac{n+1}{x'_i}\right)\right]\frac{\rho^2}{2}$$

we substitute $-\frac{1}{x_o}+\frac{1}{x_i}=(n-1)\left(\frac{1}{r_1}-\frac{1}{r_2}\right)$

$$\left(\frac{1}{r_2}-\frac{1}{x'_i}\right)^2=\left(\frac{1}{r_2}-\frac{1}{x_o}-(n-1)\left(\frac{1}{r_1}-\frac{1}{r_2}\right)\right)^2=\left(-\frac{n}{r_2}+\frac{1}{x_o}+\frac{n-1}{r_2}\right)^2$$

and $$\left(\frac{1}{r_2}-\frac{n+1}{x'_i}\right)=\frac{1}{r_2}-(n+1)\left[\frac{1}{x_o}+(n-1)\left(\frac{1}{r_1}+\frac{1}{r_2}\right)\right]=\left(-\frac{n+1}{x_o}-\frac{n^2-1}{r_1}+\frac{n^2}{r_2}\right)$$

(c) We obtain:

$$-\frac{1}{x_o}+\frac{1}{x'_i}=(n-1)\left(\frac{1}{r_1}-\frac{1}{r_2}\right)+\frac{n-1}{n^2}\left[\left(\frac{1}{r_1}-\frac{1}{x_o}\right)^2\left(\frac{1}{r_1}-\frac{n+1}{x_o}\right)\right.$$

$$\left.-\left(\frac{1}{x_o}+\frac{n-1}{r_1}-\frac{n}{r_2}\right)^2\left(\frac{n^2}{r_2}-\frac{n+1}{x_o}-\frac{n^2-1}{r_1}\right)\right]\frac{\rho^2}{2}$$

3. *The parameters π and σ.* We introduce the parameters

$$\pi=\frac{x'_i+x_o}{x'_i-x_o}$$

and

$$\sigma=\frac{r_2+r_1}{r_2-r_1}$$

a. Using

$$\frac{1}{x'_i}-\frac{1}{x_o}=\frac{1}{f}=(n-1)\left(\frac{1}{r_1}-\frac{1}{r_2}\right)$$

find the expressions

$$\frac{1}{x_o}=-\frac{\pi+1}{2f},\quad \frac{1}{x_i}=\frac{1-\pi}{2f}$$

$$\frac{1}{r_1}=\frac{\sigma+1}{2f(n-1)},\quad \frac{1}{r_2}=\frac{\sigma-1}{2f(n-1)}$$

b. Introducing the parameters π and σ into the result of problem 2, show that we finally get

$$-\frac{1}{x_o}+\frac{1}{x'_i}=(n-1)\left(\frac{1}{r_1}-\frac{1}{r_2}\right)+\frac{\rho^2}{f^3}(A\sigma^2+B\sigma\pi+C\pi^2+D)$$

where

$$A = \frac{n+2}{8n(n-1)^2}, \quad B = \frac{n+1}{2n(n-1)}$$
$$C = \frac{3n+2}{8n}, \quad D = \frac{n^2}{8(n-1)^2}$$

Solution

(a)

$$\frac{1}{x_i'} - \frac{1}{x_o} = \frac{1}{f} \qquad (x_i' - x_o)\pi = x_i' + x_o$$

$$x_i' = \frac{x_o\ (\pi + 1)}{\pi - 1}$$

$$\frac{1}{x_o} = -\frac{1}{f} + \frac{\pi - 1}{x_o(\pi + 1)}$$

$$1 - \frac{\pi - 1}{\pi + 1} = -\frac{x_o}{f} \qquad \rightarrow \quad \frac{1}{x_o} = -\frac{\pi + 1}{2f}$$

Similarly we get

$$\frac{1}{x_i'} = \frac{1}{x_o} + \frac{1}{f} = -\frac{\pi + 1}{2f} + \frac{1}{f} = -\frac{\pi + 1}{2f}$$

$$\frac{1}{x_i'} = \frac{1 - \pi}{2f}$$

$$\frac{1}{f} = (n-1)(\frac{1}{r_1} - \frac{1}{r_2}) \qquad \sigma = \frac{r_2 + r_1}{r_2 - r_1} \quad \rightarrow \quad r_2 = r_1\frac{\sigma + 1}{\sigma - 1}$$

$$\frac{1}{r_1} = \frac{1}{r_2} + \frac{1}{f(n-1)}$$

$$\frac{1}{r_1} = \frac{1}{r_1}(\frac{\sigma-1}{\sigma+1}) + \frac{1}{f(n-1)} \qquad \text{Thus} \qquad \frac{1}{r_1} = \frac{\sigma + 1}{2f(n-1)}$$

Substituting for r_1 we get $\frac{1}{r_2} = \frac{\sigma - 1}{2f(n-1)}$

(b) We substitute

$$\frac{1}{x_o} = -\frac{\pi+1}{2f} \qquad \frac{1}{r_1} = \frac{\sigma + 1}{2f(n-1)} \qquad \frac{1}{x_i'} = \frac{1-\pi}{2f} \qquad \frac{1}{r_2} = \frac{\sigma - 1}{2f(n-1)}$$

into the result of problem 2.

$$\frac{1}{x_o} + \frac{1}{x_i} = (n - 1)\ (\frac{1}{r_1} - \frac{1}{r_2}) +$$

$$\underbrace{\frac{n-1}{n^2}\left[\left(\frac{\sigma+1}{2f(n-1)}+\frac{\pi+1}{2f}\right)^2\left(\frac{\sigma+1}{2f(n-1)}+\frac{n+1}{2f}(\pi+1)\right)-\left(\frac{\sigma-1}{2f(n-1)}-\frac{1-\pi}{2f}\right)^2\left(\frac{\sigma-1}{2f(n-1)}-(n+1)\frac{(1-\pi)}{2f}\right)\right]}_{\text{We let this part } = Y}\frac{\rho^2}{2}$$

$$Y = \frac{n-1}{n^2 8f^3(n-1)^3}\left[\frac{\sigma+1}{1}+\frac{(\pi+1)}{1}(n-1)^2(\sigma+1+(n+1)(n-1)(n+1)\right.$$

$$\left.-(\sigma-1-(1-\pi)(n-1))^2(\sigma-1-(n+1)(n-1)(1-\pi))\right]$$

$$Y\frac{\rho^2}{2} = \frac{\rho^2}{n^2\, 16\, f^3(n-1)^2}\{(\pi(1-n)-n-\sigma)^2(\pi(n^2-1)+n^2+\sigma) -$$

$$(\pi(1-n)+n-\sigma)^2(\pi(n^2-1)-n^2+\sigma)\}$$

After simplifications:

$$\{\} = -4\pi^2(1-n)n(n^2-1)+2n^4+2n^2\sigma^2+4n\sigma^2+2\pi^2(1-n^2)n^2$$

$$-4\pi(1-n)(n^2+n+(n+1)n)\sigma$$

$$\frac{2n^4\rho^2}{n^2 16f^2(n-1)^2} = \frac{\rho^2}{f^3}\left(\frac{n^2}{8(n-1)^2}\right) \to D$$

$$\pi^2 \text{ term} \quad \frac{\rho^2\pi^2}{n^2 16f^3(n-1)^2}\left(4(1-n)n(n^2-1)+2(1-n)^2n^2\right)$$

$$= \frac{\pi^2\rho^2}{f^3}\left(\frac{3n+2}{8n}\right) \to C$$

$$\pi\ \sigma \text{ term:} \quad \frac{\pi\cdot\sigma(-4(1-n)(n^2+n+n^2+n))\cdot\rho^2}{n^2\cdot 16(n-1)^2 \qquad f^3}$$

$$= \frac{\rho^2}{f^3}\left(\frac{n+1}{2n(n-1)}\right) \to B$$

$$\sigma^2 \text{ term:} \quad \frac{\rho^2\sigma^2}{f^3}\ \frac{(2n^2+4n)}{n^2\cdot 16\cdot(n-1)^2} = \frac{\rho^2\sigma^2}{f^3}\left(\frac{(n+2)}{8n(n-1)^2}\right) \to A$$

4. *Thin lens and spherical aberration.* Can the spherical aberration be corrected for a thin lens? If the expression

$$Y = A\sigma^2 + B\sigma\pi + C\pi^2 + D$$

could be equal to zero for specific values of σ, π, and n, then we would have no spherical aberration. Y can be plotted as a function of σ and will give a family of parabolas depending on n and π as parameters.

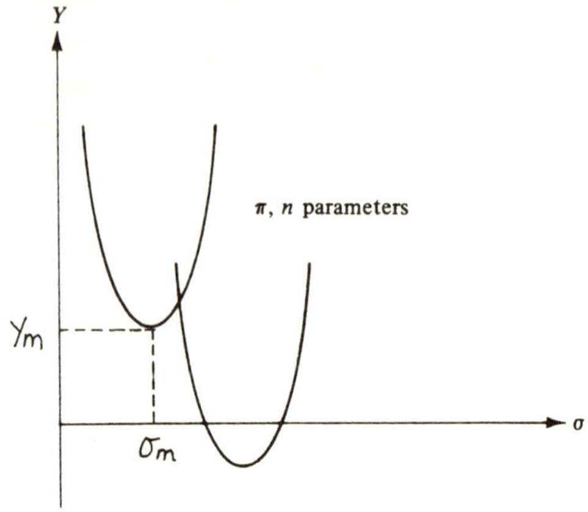

Now we want to show that $Y = 0$ is not possible for any value of σ if $n = 1.5$.

a. Calculate A, B, C, and D for $n = 1.5$.

b. Show that the minimum value of $Y(\sigma)$ is for

$$\sigma_m = -\frac{B\pi}{2A} = -0.71\pi$$

c. Show that

$$Y(\sigma_m) = 4.464\sigma_m^2 + 1.125$$

and that, consequently, there is no real solution for σ_m.

We see that for a single thin lens, spherical aberration cannot be eliminated for any variation of x_o, x_i, r_1, or r_2, given $n = 1.5$. Nor is it possible to eliminate aberration for other practical values of n.

Solution

(a)

A = 1.167 B = 1.667

C = .542 D = 1.125

(b)

$$A\sigma^2 + B\sigma\pi + C\pi^2 + D = 0$$

$$\sigma^2 + 1.428\sigma\pi + .464\pi^2 + .964 = 0$$

The parabola $Y(\sigma) = 1.167\sigma^2 + 1.667\pi + .542\pi^2 + 1.125$

has its min. value at

$$\sigma_m = -\frac{B\pi}{2A} = -.71\pi$$

(c)

$$Y(\sigma_m) = 1.167\sigma_m^2 + 2.234\sigma_m^2 + 1.063\sigma_m^2 + 1.125$$

$$= 4.464\sigma_m^2 + 1.125$$

No solution for Y = 0.

5. *Coma.* In a similar way as discussed for spherical aberration, we can obtain an expression for the distance Q'_0 to Q'_4 (see Figure 13.13). Q'_0 refers to the image point of the chief ray; Q'_4 refers to the image point of the two outermost rays. This expression is

$$Q'_0Q'_4 = \frac{3\rho^2 \tan\beta}{2fn(n-1)}[(n+1)\sigma + (2n+1)(n-1)\pi]\frac{1}{1-\sigma}$$

where

$$\pi = \frac{x_i + x_o}{x_i - x_o} \quad \text{and} \quad \sigma = \frac{r_2 + r_1}{r_2 - r_1}$$

and where β is the angle made by the chief ray with the axis, and where ρ is the distance of the zone being considered from the axis, and f is the focal length of the lens.

a. Show that coma is eliminated if the condition

$$\sigma = -(2n+1)\frac{n-1}{n+1}\pi$$

is met.

b. Calculate σ for the case where the object is at infinite distance and $n = 1.5$.

c. What is the percentage difference of the values of σ for no coma and σ for minimum spherical aberration for $\pi = -1$?

Solution

(a) The bracket containing the position factor π and shape factor σ must vanish

$$(n+1)\sigma + (2n+1)(n-1)\pi = 0$$

$$\sigma = -(2n+1)\frac{n-1}{n+1}\pi$$

(b) Object at infinite distance: $\pi = -1$

$$\sigma = (2n+1)\frac{n-1}{n+1}$$

$$n = 1.5$$

$$\sigma = 4\,\frac{0.5}{2.5} = \frac{2}{2.5} = 0.8$$

(c) $\sigma_{\text{min. sph. ab.}} = 0.71 \qquad \sigma_{\text{no coma}} = 0.8$

$$\frac{0.8 - 0.71}{0.8} 100 = 11.25\%$$

We see that a lens designed for minimum spherical aberration is close to the condition of "no coma".

6. *Aplanatic lens.* A lens free of spherical aberration and coma is called an aplanatic lens. For such a lens, the sine condition is fulfilled. We know that spherical aberration can only be eliminated if object and image are not both real. For the elimination of coma, we found that

$$\sigma = -\frac{(2n+1)(n-1)}{n+1}\pi$$

a. Show that the two values

$$\sigma = -(2n+1) \quad \text{and} \quad \pi = \frac{n+1}{n-1}$$

make the expressions for spherical aberration and coma both equal to zero.

b. Introduce these values of σ and π into the following expressions discussed in problem 3:

$$\frac{1}{x_o} = -\frac{\pi+1}{2f}, \quad \frac{1}{x_i} = \frac{1-\pi}{2f}$$

$$\frac{1}{r_2} = \frac{\sigma-1}{2(n-1)f}, \quad \frac{1}{r_1} = \frac{\sigma+1}{2(n-1)f}$$

Derive the following relations:

$$f = \frac{n+1}{1-n}r_2, \quad x_o = \frac{n+1}{n}r_2, \quad x_i = (n+1)r_2, \quad r_1 = \frac{n+1}{n}r_2$$

from which we get

$$r_2 = \frac{n}{n+1}r_1, \quad x_o = r_1, \quad x_i = nr_1.$$

These are the formulas used in the discussion of the aplanatic lens (Section 2 of this chapter).

c. Consider an aplanatic lens of refractive index $n = 1.5$ and radius of curvature $r_1 = -5$ cm. Calculate r_2, x_o, x_i, and f, and confirm that the thin-lens equation is fulfilled.

d. Make a sketch of the positions of x_o, x_i, and r_2.

<u>Solution</u>

(a) The two expressions to be considered are

$$\sigma = -\frac{(2n+1)(n-1)}{n+1}\pi$$

and

$$\frac{n+2}{8n(n-1)^2}\sigma^2 + \frac{n+1}{2n(n-1)}\sigma\pi + \frac{3n+2}{8n}\pi^2 + \frac{n^2}{8(n-1)^2}$$

The first is fulfilled with $\sigma = -(2n+1)$ and

$\pi = \frac{n+1}{n-1}$. Introducing these values into the second expression and equating it to zero yields:

$$\frac{n+2}{8n(n-1)^2}(2n+1)^2 + \frac{n+1}{2n(n-1)}(-2n-1)\frac{n+1}{n-1} + \frac{3n+2}{8n}\frac{(n+1)^2}{(n-1)^2}$$

$$+ \frac{n^2}{8(n-1)^2} = 0$$

$$\frac{(n+2)(2n+1)^2}{8n} + \frac{(n+1)^2(-2n-1)}{2n} + \frac{(3n+2)}{8n}(n+1)^2 + \frac{n^2}{8} = 0$$

$$(n+2)(2n+1)^2 + 4(n+1)^2(-2n-1) + (3n+2)(n+1)^2 + n^3 = 0$$

$$(n+2)(4n^2+4n+1) + (n^2+2n+1)(-8n-4) + (3n+2)(n^2+2n+1) + n^3 = 0$$

$$4n^3 + 4n^2 + n + 8n^2 + 8n + 2$$

$$-8n^3 - 16n^2 - 8n - 4n^2 - 8n - 4$$

$$+3n^3 + 6n^2 + 3n + 2n^2 + 4n + 2$$

$$+n^3 = 0$$

<u>Solution</u>

(b)

$$\frac{1}{x_o} = -\frac{\pi+1}{2f} = -\frac{\frac{n+1}{n-1}+1}{2f} = -\frac{n}{f(n-1)}$$

$$\frac{1}{x_i} = \frac{1-\frac{n+1}{n-1}}{2f} = \frac{\frac{n-1-(n+1)}{n-1}}{2f} = \frac{-1}{f(n-1)}$$

$$\frac{1}{r_2} = \frac{-(2n+1)-1}{2(n-1)f} = \frac{-2n-2}{2(n-1)f} = -\frac{n+1}{(n-1)f}$$

$$\frac{1}{r_1} = \frac{-2n-1+1}{2(n-1)f} = \frac{-2n}{2(n-1)f} = \frac{-n}{(n-1)f}$$

$$f = \frac{n+1}{1-n}\ r_2$$

$$x_o = -f\,\frac{n-1}{n} = \frac{n+1}{n}\ r_2$$

$$x_i = -(n-1)f = (n+1)\ r_2$$

$$r_1 = -\frac{n-1}{n}\,f = \frac{n+1}{n}\,r_2$$

$$r_2 = \frac{n}{n+1}\,r_1\,, \quad x_o = r_1\,, \quad x_i = nr_1$$

(c)

$$r_2 = \frac{n}{n+1} r_1 = \frac{1.5}{2.5} (-5cm) = -3cm$$

$$x_o = \frac{n+1}{n} (-3) = \frac{2.5}{1.5} (-3) = -5cm$$

$$x_i = (n+1)(-3) = 2.5(-3) = -7.5cm$$

$$f = \frac{n+1}{1-n} (-3) = \frac{2.5}{0.5} (-3) = 15cm$$

The thin lens equation

$$-\frac{1}{x_o} + \frac{1}{x_i} = \frac{1}{f}$$

$$\frac{1}{5} - \frac{1}{7.5} = \frac{1}{15}$$

(d)

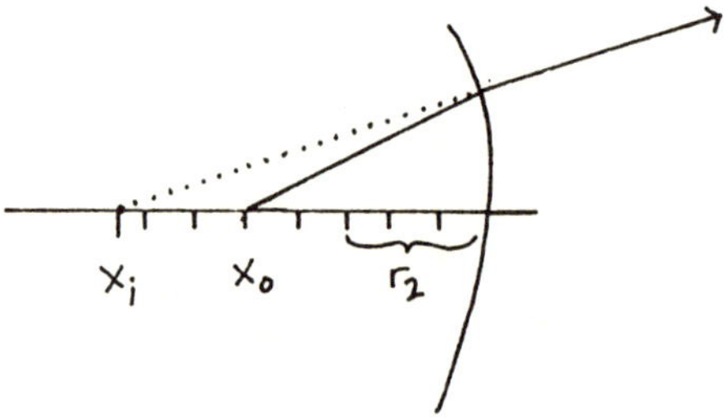

7. *Aplanatic sphere.* Let us now treat the aplanatic sphere with the imaging equation at one surface. We have from Chapter 1:

$$-\frac{n}{x_o} + \frac{n'}{x_i} = \frac{n' - n}{r}$$

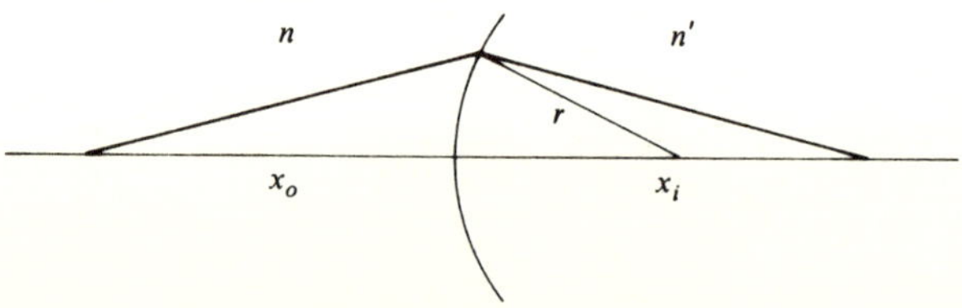

a. Write the imaging equation for one surface of a glass sphere, where $n = 1.5$ and $n' = 1$.

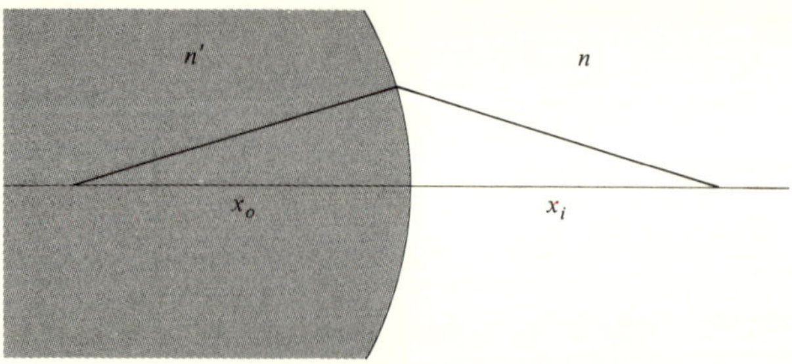

b. Assume for

$$x_0 = -\frac{n+1}{n} r$$

(we set $r_2 = -r$) that is the aplanatic distance of point P at x_o from the curved surface. Show that we get $x_i = nx_o$, that is the aplanatic distance for $x_i = P'$ from the curved surface.

Solution

(a) with n=1.5

$$-\frac{n}{x_o} + \frac{1}{x_i} = \frac{1-n}{-r} = \frac{n-1}{r} \qquad -\frac{1.5}{x_o} + \frac{1}{x_i} = \frac{0.5}{r}$$

(b)

$$-\frac{n}{-\frac{n+1}{n} r} + \frac{1}{x_i} = \frac{n-1}{r}$$

$$\frac{n^2}{n+1} + \frac{r}{x_i} = n-1$$

$$\frac{r}{x_i} = n-1 - \frac{n^2}{n+1} = \frac{(n-1)(n+1)-n^2}{n+1}$$

$$= \frac{n^2-1-n^2}{n+1} = \frac{-1}{n+1}$$

$$x_i = -(n+1)r$$

to compare with the formulas of the problem "aplanatic lens" we have to set $-r = r_2$

8. *Astigmatism at a single surface.* We first consider the focal points in the horizontal plane (sagittal) (see Figure 13.17 and following figure). We measure x_o along the line

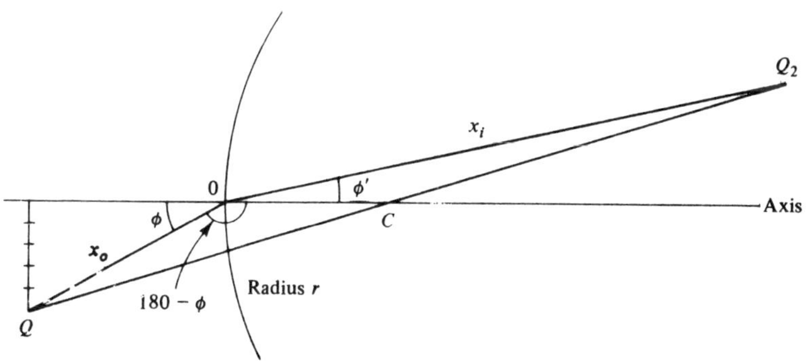

$Q0$ and x_i along the line $0Q_2$ (as in the paraxial case). For the triangles, we get

$$Q0C = \tfrac{1}{2}r|x_o|\sin\phi$$
$$0CQ_2 = \tfrac{1}{2}rx_i\sin\phi'$$
$$Q0C + 0CQ_2 = Q0Q_2 = \tfrac{1}{2}|x_o|x_i\sin(180 - \phi + \phi')$$

or

$$r|x_o|\sin\phi + rx_i\sin\phi' = |x_o|x_i\sin(\phi - \phi')$$

a. Apply

$$\sin(\alpha + \beta) = \sin\alpha\cos\beta + \cos\alpha\sin\beta$$

and

$$\sin\phi = n\sin\phi'$$

and obtain

$$-\frac{1}{x_o} + \frac{n}{x_i} = \frac{n\cos\phi' - \cos\phi}{r}$$

where at the end we set $|x_o| = -x_o$ according to our sign convention. We now consider the vertical (or meridional) focal points (see Figure 13.17 and following figure).

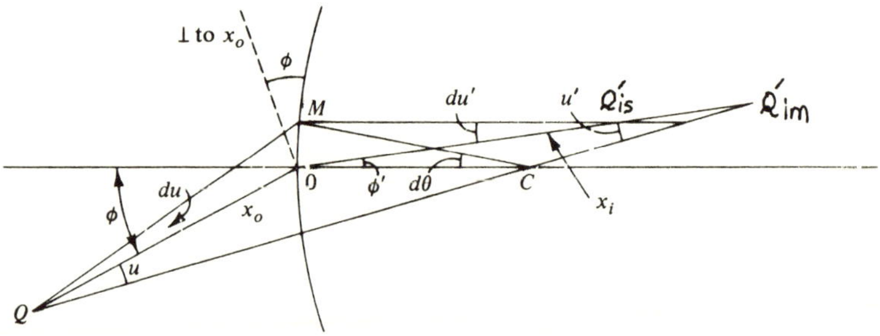

For small angles, we have

$$du = \frac{OM\cos\phi}{|x_o|}, \quad |d\theta| = \frac{OM}{r}, \quad |du'| = \frac{OM\cos\phi'}{x_i}$$

We differentiate $\sin\phi = n\sin\phi'$; that is,

$$\cos\phi\, d\phi = n\cos\phi'\, d\phi'$$

We have

$$\phi = u + |\theta|, \quad \phi' = |\theta| - |u'|$$

that is,

$$(du + |d\theta|)\cos\phi = n(|d\theta| - du')\cos\phi'$$

b. Combine the preceding relations and arrive at

$$-\frac{\cos^2\phi}{x_o} + \frac{n\cos^2\phi'}{x_i} = \frac{n\cos\phi' - \cos\phi}{r}$$

When $\phi = \phi' = 0$, both focal points Q_1 and Q_2 arc determined by the same equation

$$-\frac{1}{x_o} + \frac{n}{x_i} = \frac{n-1}{r}$$

We choose $x_o = -40.000$ cm, $r = 10.000$ cm, and $n = 1.5000$

c. Find x_i.

d. For $r = 10.000$ cm, $x_o = -40.000$ cm, $n = 1.5000$, and $\phi = 3°$, calculate the values of x_{is} from (a) and x_{im} from (b), where s and m stand for sagittal and meridional, respectively, and give the difference $x_{is} - x_{im}$ in centimeters.

Solution

(a)

$$r|x_o|\sin\phi + r\, x_i \sin\phi' = |x_o|x_i\, [\sin\phi \cos\phi' - \cos\phi \sin\phi']$$

$$r|x_o|\sin\phi + rx_i \frac{\sin\phi}{n} = |x_o|x_i\, [\sin\phi \cos\phi' - \cos\phi \frac{\sin\phi}{n}]$$

$$r|x_o| + \frac{rx_i}{n} = |x_o|x_i \cos\phi' - |x_o|x_i \frac{\cos\phi}{n}$$

$$\frac{nr}{x_i} + \frac{r}{|x_o|} = n \cos\phi' - \cos\phi$$

$$-\frac{1}{x_o} + \frac{n}{x_i} = \frac{n \cos\phi' - \cos\phi}{r}$$

(b)

$$(du + |d\theta|)\cos\phi = n\, (|d\theta| - |du'|)\cos\phi'$$

$$OM\, (\frac{\cos\phi}{|x_o|} + \frac{1}{r}) \cos\phi = n\, (\frac{1}{r} - \frac{\cos\phi'}{x_i})\, OM \cos\phi'$$

$$\frac{\cos^2\phi}{|x_o|} + \frac{\cos\phi}{r} = \frac{n \cos\phi'}{r} - \frac{n \cos^2\phi'}{x_i}$$

$$-\frac{\cos^2\phi}{x_o} + \frac{n \cos^2\phi'}{x_i} = \frac{n \cos\phi' - \cos\phi}{r}$$

(c)

$$\frac{1}{40.000} + \frac{1.5000}{x_i} = \frac{0.5000}{10.000}$$

$$x_i = 60.000 \text{ cm.}$$

(d)

Sagittal

$$\frac{n}{x_{is}} = \frac{n \cos \phi' - \cos \phi}{10.000} - \frac{1}{40.000}$$

$\sin \phi = n \sin \phi'$ $\quad \phi = 3°$ $\quad \phi' = 1.9995°$

$\cos \phi = 0.99863$ $\quad \cos \phi' = 0.99939$

$$\frac{1.5}{x_{is}} = \frac{1.4991 - 0.99939}{10.000} - 0.025$$

$$= 0.049969 - 0.025 = 0.02497$$

$$x_{is} = 60.072$$

Meridional

$$\frac{n \cos^2 \phi'}{x_{im}} = \frac{n \cos \phi' - \cos \phi}{10.000} - \frac{\cos^2 \phi}{40.000}$$

$$\frac{n \cdot 0.99878}{x_{im}} = \frac{1.5 \cdot 0.99939 - 0.99863}{10.000} - \frac{0.99726}{40.000}$$

$$= 0.050046 \qquad - 0.024932$$

$$= 0.025114$$

$$x_{im} = 59.655\text{cm}$$

$$x_{is} - x_{im} = .417\text{cm}$$

9. *Achromatic doublet with* $t = 0$. In Chapter 1, we obtained the following focal lengths for two thin lenses in contact:

$$\frac{1}{f} = \frac{1}{f_1} + \frac{1}{f_2} = (n_1 - 1)\left(\frac{1}{r_1} - \frac{1}{r_2}\right) + (n_2 - 1)\left(\frac{1}{r_1'} - \frac{1}{r_2'}\right)$$

where n_1, r_1, r_2 and n_2, r_1', r_2' are as indicated:

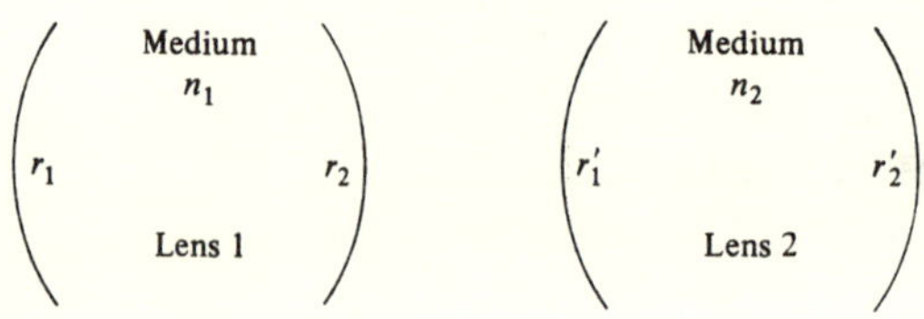

We abbreviate

$$\frac{1}{r_1} - \frac{1}{r_2} = k_1 \quad \text{and} \quad \frac{1}{r_1'} - \frac{1}{r_2'} = k_2$$

then we have

$$\frac{1}{f} = (n_1 - 1)k_1 + (n_2 - 1)k_2$$

The focal length should be independent of the refractive index in the wavelength region λ_{blue} to λ_{red}. The condition for $1/f$ is

$$\frac{d}{d\lambda}\left(\frac{1}{f}\right) = \frac{d}{d\lambda}[(n_1 - 1)k_1 + (n_2 - 1)k_2] = 0$$

or

$$\frac{dn_1}{d\lambda}k_1 + \frac{dn_2}{d\lambda}k_2 = 0$$

or

$$\frac{k_1}{k_2} = -\frac{dn_2}{dn_1}$$

We approximate

$$dn_1 = n_{1B} - n_{1R} \quad \text{and} \quad dn_2 = n_{2B} - n_{2R}$$

where n_B stands for the refractive index of blue light and n_R for the refractive index of red light. We also introduce the medium refractive index n_D, where D stands for the sodium line. We use the medium refractive index to relate k_1 and f_1 and k_2 and f_2

$$k_1 = \frac{1}{f_1(n_1 - 1)} = \frac{1}{f_1(n_{1D} - 1)}$$

$$k_2 = \frac{1}{f_2(n_2 - 1)} = \frac{1}{f_2(n_{2D} - 1)}$$

We define

$$V_1 = \frac{n_{1B} - n_{1R}}{n_{1D} - 1} \quad \text{and} \quad V_2 = \frac{n_{2B} - n_{2R}}{n_{2D} - 1}$$

a. Show that we have

$$\frac{f_2}{f_1} = -\frac{V_2}{V_1}$$

b. Assume the following values:

$$n_{1D} = 1.700 \quad n_{2D} = 1.500$$
$$n_{1B} = 1.735 \quad n_{2B} = 1.525$$
$$n_{1R} = 1.665 \quad n_{2R} = 1.475$$

What follows for f_1 and f_2, and what type of lenses must be used?

c. The resulting lens combination of problem (b) does not act as a lens. What condition must be placed on V_1 and V_2 to have a converging lens?

Solution

(a)

$$\frac{k_1}{k_2} = \frac{\dfrac{1}{f_1(n_{1D}-1)}}{\dfrac{1}{f_2(n_{2D}-1)}} = -\frac{n_{2B} - n_{2R}}{n_{1B} - n_{1R}}$$

$$\frac{f_2(n_{2D}-1)}{f_1(n_{1D}-1)} = -\frac{n_{2B}-n_{2R}}{n_{1B}-n_{1R}}$$

$$\frac{f_2}{f_1} = -\frac{\dfrac{n_{2B}-n_{2R}}{n_{2D}-1}}{\dfrac{n_{1B}-n_{1R}}{n_{1D}-1}} = -\frac{V_2}{V_1}$$

(b)

$$V_1 = \frac{1.735-1.665}{0.700} = \frac{0.07}{0.700} = .1$$

$$V_2 = \frac{1.525-1.475}{0.500} = \frac{0.05}{0.50} = .1$$

$$\frac{f_2}{f_1} = -1 \quad \text{or } f_2 = -f_1$$

The lenses must have equal focal length, one converging, one diverging.

(c) If lens 1 is convergent, and lens 2 divergent, one must have $|f_1| > |f_2|$ or $V_2 > V_1$.

10. *Achromatic doublet with $t \neq 0$.* For two thin lenses at a distance t apart, we have the following focal length according to Chapter 1:

$$\frac{1}{f} = \frac{1}{f_1} + \frac{1}{f_2} - t\left(\frac{1}{f_1 f_2}\right)$$

For one lens, we have

$$\frac{1}{f} = (n-1)\left(\frac{1}{r_1} - \frac{1}{r_2}\right)$$

Differentiation with respect to n yields

$$\Delta\left(\frac{1}{f}\right) = \Delta n\left(\frac{1}{r_1} - \frac{1}{r_2}\right) = \frac{\Delta n}{f(n-1)}$$

In problem 9, we defined

$$V_1 = \frac{n_{1B} - n_{1R}}{n_{1D} - 1}$$

which is now written as

$$V = \frac{\Delta n}{n-1}$$

and we have

$$\Delta\left(\frac{1}{f}\right) = \frac{V}{f}$$

Differentiation of $1/f$ for the two-lens system gives us

$$\Delta\left(\frac{1}{f}\right) = \frac{V_1}{f_1} + \frac{V_2}{f_2} - t\left[\frac{1}{f_2}\Delta\left(\frac{1}{f_1}\right) + \frac{1}{f_1}\Delta\left(\frac{1}{f_2}\right)\right]$$
$$= \frac{V_1}{f_1} + \frac{V_2}{f_2} - t\left(\frac{V_1}{f_1 f_2} + \frac{V_2}{f_1 f_2}\right)$$

a. Set

$$\Delta\left(\frac{1}{f}\right) = 0$$

and show that we require the following condition for no chromatic aberration:

$$t = \frac{V_1 f_2 + V_2 f_1}{V_2 + V_1}$$

b. Show that the condition for no chromatic aberration with two lenses of equal material is that they are separated by a distance of

$$t = \frac{f_1 + f_2}{2}$$

Solution

(a)

$$0 = \frac{V_1}{f_1} + \frac{V_2}{f_2} - t\left(\frac{V_1 + V_2}{f_1 f_2}\right)$$

$$V_1 f_2 + V_2 f_1 = t(V_1 + V_2)$$

$$t = \frac{V_1 f_2 + V_2 f_1}{V_1 + V_2}$$

(b)

$$V_1 = V_2$$

$$t = \frac{f_1 + f_2}{2}$$

14 Detectors

1. *Finding the amplitude of the trial solution θ_f (equation 14.5).* The differential equation

$$M\ddot{\theta} + D\dot{\theta} + G\theta = A_f e^{i2\pi ft}$$

is solved with the trial solution

$$\theta = \theta_f e^{i2\pi ft}$$

Find θ_f as a function of A_f, G, M, D, and ω.

<u>Solution</u>

$$\dot{\theta} = i2\pi f\theta_f \; e^{i2\pi ft}$$

$$\ddot{\theta} = -(2\pi f)^2 \; \theta_f \; e^{i2\pi ft}$$

$$-M(2\pi f)^2 \; \theta_f \; e^{i2\pi ft} + i2\pi f \; \theta_f \; e^{i2\pi ft} D + G \; \theta_f \; e^{i2\pi ft} = A_f e^{i2\pi ft}$$

set $2\pi f = \omega$

$$\theta_f \; (-M\omega^2 + i\omega D + G) = A_f$$

$$\theta_f = \frac{A_f}{(G - M\omega^2) + i\omega D}$$

2. *Damped oscillator.* Consider the value obtained in problem 1,

$$\frac{A_f}{(G - M\omega^2) + i\omega D}$$

Give the absolute value and the absolute value squared.

<u>Solution</u>

We abbreviate:

$$z = \frac{1}{a + ib} = \frac{a - ib}{(a + ib)(a - ib)} = \frac{a - ib}{a^2 + b^2}$$

The absolute value is obtained

$$|z| = \sqrt{zz^*} = \sqrt{\frac{(a - ib)}{a^2 + b^2} \frac{(a + ib)}{a^2 + b^2}} = \sqrt{\frac{a^2 + b^2}{((a^2 + b^2))^2}}$$

$$= \frac{1}{\sqrt{a^2 + b^2}}$$

Application to our problem yields

$$|\theta_f| = \frac{A_f}{\sqrt{(G - M\omega^2)^2 + (\omega D)^2}}$$

The square of the absolute value is

$$z^2 = \frac{1}{a^2 + b^2} \rightarrow \theta_f^2 = \frac{A_f^2}{(G - M\omega^2)^2 + (\omega D)^2}$$

Time averaging is done for θ_f^2 and $A_f^2 \rightarrow \overline{\theta_f^2}$, $\overline{A_f^2}$

3. *The R-L-C circuit.* The total power of the *R-L-C* circuit is given as

$$Iu = \frac{d}{dt}(\text{magnetic energy}) + \frac{d}{dt}(\text{electric energy}) + \text{losses}$$

We have I current, $\frac{1}{2}LI^2$ magnetic energy, L inductance, RI losses, u voltage, $\frac{1}{2}(q^2/C)$ electrical energy, C capacitance, q charge, and R resistance.

a. Get

$$\dot{u} = L\frac{d^2I}{dt^2} + \frac{1}{C}I + R\frac{dI}{dt}$$

by differentiation of the electrical and magnetical energy.

b. Use the trial solution $u = u_f e^{i\omega t}$ and $I = I_f e^{i\omega t}$ and obtain

$$\frac{I_f}{i\omega C} = \frac{u_f}{(1 - LC\omega^2) + i\omega RC}$$

Solution

(a)
$$Iu = \frac{d}{dt}\left(\frac{1}{2}LI^2\right) + \frac{d}{dt}\left(\frac{1}{2C}q^2\right) + RI^2$$

$$Iu = LI\frac{dI}{dt} + \frac{1}{C}qI + RI^2$$

$$u = L\frac{dI}{dt} + \frac{1}{C}q + RI$$

$$\dot{u} = L\frac{d^2I}{dt^2} + \frac{1}{C}I + R\frac{dI}{dt}$$

(b)

$$u = u_f e^{i\omega t}, \qquad I = I_f e^{i\omega t}$$

$$u_f(i\omega) e^{i\omega t} = L(i\omega)^2 I_f e^{i\omega t} + \frac{I_f}{C} e^{i\omega t} + RI_f(i\omega) e^{i\omega t}$$

$$\frac{u_f}{-CL\omega^2 + 1 + RCi\omega} = \frac{I_f}{i\omega C}$$

4. *Average over the noise frequency spectrum of* $\overline{v_f^2}$. Go back to equation 14.17:

$$\overline{v_f^2} = \frac{\overline{u_f^2}}{(1 - LC\omega^2)^2 + \omega^2(RC)^2}$$

Before averaging over all frequencies from 0 to ∞, we introduce the approximation $L \to 0$ and have

$$\overline{v^2} = \frac{1}{2\pi}\int_0^\infty \frac{\overline{u_f^2}\, d\omega}{1 + \omega^2(RC)^2}$$

using the integral formula

$$\int_0^\infty \frac{dx}{1 + x^2} = \frac{\pi}{2}$$

Show that we have

$$\overline{v^2} = \frac{\overline{u_f^2}}{4RC}$$

Introduction of this result into

$$\tfrac{1}{2}C\overline{v^2} = \tfrac{1}{2}kT$$

yields Nyquist's formula,

$$\overline{u_f^2} = 4RkT$$

<u>Solution</u>

$$\overline{v^2} = \overline{u_f^2}\,\frac{1}{2\pi}\int_0^\infty \frac{d\omega}{1 + \omega^2(RC)^2}$$

$$2\pi\ RC\,f = x \qquad df = \frac{dx}{2\pi RC}$$

$$\overline{v^2} = \overline{u_f^2}\ \frac{1}{2\pi RC}\int_0^\infty \frac{dx}{1 + x^2} = \frac{\pi}{2}\ \frac{1}{2\pi RC}\ \overline{u_f^2}$$

$$\overline{v^2} = \frac{1}{4RC}\ \overline{u_f^2}$$

5. *Numerical calculation of* $\overline{v^2}$. Calculate

$$\sqrt{\overline{v^2}} = \sqrt{4RkT\,df}$$

for $T = 2$ K and $R = 1 \times 10^6\ \Omega$.

Solution

$$\sqrt{\overline{v^2}} = \sqrt{(4)(10^6)(1.38 \times 10^{-23})\ 2 \cdot \frac{\Omega J}{K}\ 1 \cdot \frac{1}{s}}$$

we have $\frac{\Omega J}{s} = \frac{V}{A} W = \frac{V}{A} AV = V^2$

$$\sqrt{\overline{v^2}} = \sqrt{8(1.38) \times 10^{-17}}\ V$$

$$= \sqrt{11.04 \times 10^{-17}} = \sqrt{1.10 \times 10^{-16}}\ V = 1 \times 10^{-8}\ \text{Volt}$$

6. *Calculation of* Δv. Consider the circuit in Figure 14.5. The ratio of the voltage drop across R and $R_L + R$ is

$$\frac{v_D}{V} = \frac{R}{R_L + R}$$

Calculate the change in voltage Δv depending on the change in the resistor ΔR.

Solution

$$v_D = \frac{R \cdot V}{R_L + R}$$

$$\frac{dv_D}{dR} = \frac{(R_L + R)V - RV}{(R_L + R)^2}$$

$$\Delta v_D = \frac{R_L V}{(R_L + R)^2}\ \Delta R$$

7. *Responsivity of a bolometer*. Calculate the responsivity

$$r = \frac{\Delta v}{\Delta W} = \alpha I R \tilde{R}$$

for the semiconducting bolometer using $\alpha = -0.03$ 1/K and $R = 10^6\ \Omega$.
a. Calculate $\tilde{R}$ by using $\tilde{G}_0 = 10^{-4}$ W/K and $T - T_0 = 10$ K (equation 14.55).

b. Calculate I from

$$W_0 = I^2 R = \tilde{G}_0 (T - T_0)$$

c. Calculate the absolute value of the responsivity using $r = \alpha I R \tilde{R}$.

Solution

(a) $\tilde{G} = G_o [1 - \alpha (T - T_o)]$

$$\tilde{G} = 10^4 \frac{W}{K} [1 - (-.03 \frac{1}{K})(10K)]$$

$$\tilde{G} = 1.3 \times 10^{-4} \frac{W}{K}$$

$$\tilde{R} = \frac{1}{\tilde{G}} = \frac{1}{1.3 \times 10^{-4}} \frac{K}{W} = .77 \times 10^4 \frac{K}{W}$$

$$\tilde{R} = .77 \times 10^4 \frac{K}{W}$$

(b) $W_o = I^2 R = \tilde{G}_o (T - T_o)$

$$I = \sqrt{\frac{\tilde{G}_o (T - T_o)}{R}}$$

$$I = \sqrt{\frac{10^{-4} \frac{W}{K}(10K)}{10^6 \Omega}} = 3.16 \times 10^{-5} \sqrt{\frac{A^2 \Omega}{\Omega}}$$

$$I = 3.16 \times 10^{-5} A$$

(c)

$$|r| = |(-.03 \frac{1}{K})(3.16 \times 10^{-5} A)(10^6 \Omega)(.77 \times 10^4 \frac{K}{W})|$$

$$|r| = 7.3 \times 10^3 \frac{A\Omega}{W} = .73 \times 10^4 \frac{V}{W}$$

$$|r| = 0.73 \times 10^4 \frac{V}{W}$$

8. *Thermal noise and Johnson noise.* Calculate the combination of thermal noise and Johnson noise for the bolometer. Use $T = 4$ K, $T_0 = 3.999$ K, $\alpha = -(3000 \text{ K}/T^2)$ [1/K], $\tilde{G}_0 = 10^{-4}$ W/K, and $k = 1.4 \times 10^{-23}$ J/K.
   a. Calculate α.
   b. Calculate $\tilde{R}$ from $\tilde{G} = \tilde{G}_0[1 - \alpha(T - T_0)]$
   c. Calculate

$$\text{NEP} = \sqrt{\overline{W_B^2}} = \sqrt{\frac{4kT^2}{\tilde{R}} \left(1 + \frac{1}{T(T - T_0)\alpha^2}\right) \Delta f}$$

Solution

(a) $\alpha = -\dfrac{3000}{16} = -187.5 \ \dfrac{1}{K}$

(b) $\tilde{G} = 10^{-4}\ \dfrac{W}{K}\ (1 + 187(.001)) = 1.19 \times 10^{-4}\ \dfrac{W}{K}$

$\tilde{R} = 0.84 \times 10^{4}\ \dfrac{K}{W}$

(c)

$$\text{NEP} = \sqrt{\overline{W_B^{\ 2}}} = \sqrt{\frac{4kT^2}{\tilde{R}}\left(1 + \frac{1}{T(T - T_o)\alpha^2}\right)\Delta f}$$

$$= \sqrt{\frac{4 \times (1.4) \times 10^{-23} \times 4^2}{0.8 \times 10^4}\left(1 + \underbrace{\frac{1}{4(.001)(187)^2}}_{\to\ 1}\right)\frac{J}{K}\,K^2\,\frac{W}{K}\,\frac{1}{s}}$$

$$= \sqrt{112 \times 10^{-27}\ W^2} = 3.3 \times 10^{-13}\ W$$

9. *Responsivity of a photocell.* Calculate the responsivity of the photocell, using

$$r = \frac{eR_Lq}{h\nu}$$

where $\lambda = 0.8\ \mu\text{m}$, $R_L = 10^6\ \Omega$, $q = 10^{-2}$, $h = 6.6 \times 10^{-34}\ \text{J}\cdot\text{s}$, and $e = 1.6 \times 10^{-19}$ C.

Solution

$\lambda = 0.8\mu m = 0.8 \times 10^{-4}$ cm.

$$\nu = \frac{c}{\lambda} = \frac{3 \times 10^{10}\ \text{cm/s}}{.8 \times 10^{-4}\ \text{cm}} = 3.75 \times 10^{14}\ \frac{1}{s}$$

$$r = \frac{e\ R_L\ q}{h\nu}$$

$$r = \frac{(1.6 \times 10^{-19}C)(10^6\Omega)(10^{-2})}{(6.6 \times 10^{-34}J\cdot s)(3.75 \times 10^{14}\ \frac{1}{s})} = 6.46 \times 10^3\ \left(\frac{C}{s}\right)\ \frac{\Omega}{J\left(\frac{1}{s}\right)}$$

$$r = 6.46 \times 10^3 \ \frac{A\Omega}{W} = 6.46 \times 10^3 \ \frac{V}{W}$$

$$r = 6.46 \times 10^3 \ \frac{V}{W}$$

10. *Electrical bandwidth and frequency response.* If a system has a uniform frequency response in the interval $\Delta f = f_2 - f_1$ and no response at other frequencies, Δf is called the electrical bandwidth. For a more complicated case, where the frequency response is nonuniformly over the frequency range, we define the electrical bandwidth as

$$\Delta f = \int_0^\infty \left| \frac{r(f)}{r_{\max}} \right|^2 df$$

where $r(f)$ is the responsivity of the detector as a function of frequency. For example, for the RC circuit, we have

$$r(f) = \frac{r_{\max}}{\sqrt{1 + (\omega CR)^2}}$$

and we get

$$\Delta f = \int_0^\infty \frac{1}{1 + (\omega CR)^2} df$$

With our result from problem 4, we have

$$\Delta f = \frac{1}{4RC}$$

If the responsivity $r(f)$ is averaged over the sampling time T, we must calculate

$$r(f) = \frac{1}{T} \int_{-T/2}^{+T/2} e^{-i2\pi f t} \, dt$$

a. Show that we get (for the real part) $r(f) = \dfrac{\sin \pi f T}{\pi f T}$

b. Calculate Δf for this case.

<u>Solution</u>

(a)

$$r(f) = \frac{1}{T} \int_{-\frac{T}{2}}^{+\frac{T}{2}} e^{-i2\pi ft} \, dt = \frac{1}{T} \left| \frac{e^{-i2\pi ft}}{-i2\pi f} \right|_{-\frac{T}{2}}^{+\frac{T}{2}}$$

$$= \frac{1}{T} \left(\frac{e^{-i2\pi f\frac{T}{2}} - e^{+i2\pi f\frac{T}{2}}}{-i2\pi f} \right)$$

$$= \frac{e^{i\pi ft} - e^{-i\pi fT}}{2i\pi fT} = \frac{\sin\pi fT}{\pi fT}$$

(b)

$$\Delta f = \int_0^\infty \left(\frac{\sin \pi f T}{\pi f T}\right)^2 df$$

using $\int_0^\infty \frac{\sin^2 x}{x^2} dx = \frac{\pi}{2}$

we have $\Delta f = \frac{1}{2T}$

If we take for T the time constant RC, the frequency interval corresponds to what has been used for eq.(67).

11. *Shot noise.* Calculate the "shot noise" (equation 14.89) for the values $\lambda = 0.8\ \mu m$, $\bar{I}_t = 2\ \mu A$, $\Delta f = 1/s$, $q = 10^{-2}$, $e = 1.6 \times 10^{-19}$ C, and $h = 6.6 \times 10^{-34}$ J·s.

Solution

$\nu = 3.75 \times 10^{14}\ \frac{1}{s}$ See problem 9.

$$\overline{W^2} = 2\ e\ \bar{I}_t\ \frac{h^2 \nu^2}{q^2 e^2}$$

$$= \frac{2 \cdot (1.6) \times 10^{-19} \times 2 \times 10^{-6}\ (6.6)^2\ (10^{-34})^2 (3.75)^2\ 10^{+28}}{10^{-4}\ (1.6)^2\ (10^{-19})^2} \frac{CA(Js)^2\ \frac{1}{s}}{C^2}$$

$$= \frac{2 \cdot 2 \times (6.6)^2\ (3.75)^2 \times 10^{-6-68+28}}{(1.6) \times (10^{-19})\ 10^{-4}} \qquad \frac{AJ^2}{C}$$

$$= 1531 \times 10^{-6-68+28+19+4} \qquad \frac{A\ J^2\ V}{J}$$

multiplication by $\frac{1}{s}$

$$= 1531 \times 10^{-23}\ W \frac{J}{s}$$

$$\sqrt{\overline{W^2}} = 12 \times 10^{-11}\ W$$

12. *Johnson noise.* Calculate Johnson noise for the photodiode described by

$$\overline{W^2} = \frac{h^2 \nu^2}{q^2 e^2}\left(\frac{4kT}{R_L}\right)$$

Use $R_L = 10^7\ \Omega$, $\lambda = 0.8\ \mu m$, $q = 10^{-2}$, $T = 300\ K$, $e = 1.6 \times 10^{-19}\ C$, $h = 6.6 \times 10^{-34}\ J \cdot s$, and $k = 1.4 \times 10^{-23}$ J/K.

Solution

We have $\nu = 3.75 \times 10^{14}\ \frac{1}{s}$

$$\overline{W^2} = \frac{(6.6)^2\ (3.75)^2 \cdot 4(1.4) \cdot 3 \times 10^{-68}\ 10^{28}\ 10^{-23}\ 10^2\ J^2\ s^2\ \frac{1}{s^2}\ J\ K}{(1.6)^2\qquad 10^{-4}\ (10^{-19})^2\quad 10^7\quad C^2 K\Omega}$$

$$= \frac{(43.6)(14) \cdot 4\ (1.4) \cdot 3}{(2.6)}\quad 10^{-68+28-23+2+4+38-7}\ VJA$$

multiplication by $\frac{1}{s}$

$$= 3944 \times 10^{-26}\ (\text{Watt})^2$$

$$\sqrt{\overline{W^2}} = 63 \times 10^{-13}$$

$$\sqrt{\overline{W^2}} = .63 \times 10^{-11}\ \text{Watt}$$

13. Find the solution $\Delta n(t)$ for the equation

$$\frac{d}{dt}\Delta n = \gamma J - \frac{\Delta n}{J}$$

Solution

Trial: $\Delta N = \gamma J \tau\ (1 - e^{-\frac{t}{\tau}})$

$$\frac{d}{dt}\ \Delta N = \gamma J e^{-\frac{t}{\tau}}$$

$$\gamma J - \frac{\Delta N}{\tau} = \gamma J - \gamma J + \gamma J e^{-\frac{t}{\tau}} = \gamma J e^{-\frac{t}{\tau}}$$

14. *Responsivity of a photoconductive detector.* Calculate the responsivity of a photoconductive detector (equation 14.103), assuming that R_L is of the order of $R_L + R$. Use $I = 100 \cdot 10^{-6}$ A, $q = 10^{-2}$, $\ell = 10$ b, $h = 6.6 \times 10^{-34}$ J·s, $\nu = 3 \times 10^{14}$ 1/s ($\lambda = 1$ μm), and $A = 10$ mm². We take $\beta = 1 \times 10^{-14}$ Ω·s·m² to get r in the order of 10^4 V/W, as given in Smith, Jones, and Chasmar, *The Detection and Measurement of Infrared Radiation.*

Solution

$$r = \frac{100 \cdot 1 \cdot 1 \cdot 10 \quad 10^{-6}\ 10^{-14}\ 10^{-2} \quad A\Omega s m^2}{(6.6) \cdot 3 \cdot 10 \quad 10^{-34} 10^{14}\ 10^{-6} \quad Js \frac{1}{s} m^2}$$

$$= \frac{100}{(6.6) \cdot 3} \times 10^{-6-14-2+34-14+6} \quad \frac{A\ V\ s}{J\ A}$$

$$= 5 \times 10^4 \ \frac{V}{W}$$

15. *Johnson noise and G-R noise.* Compare Johnson noise with G-R noise for a photoconductive detector. For Johnson noise, we get with $R = 10^6$ Ω, $T = 300$ K, and $k = 1.4 \times 10^{-23}$ J/K the value $\sqrt{\overline{v^2}} = 1.3 \times 10^{-7}$ V. Calculate G-R noise. Take

$$\overline{(\delta v)^2} = \frac{R^2 I^2 2\tau}{\bar{N}}$$

and use $I = 100 \times 10^{-6}$ A, $\tau = 10^{-4}$ 1/s, $\bar{N} = 10^{18}$ 1/s, and R from above.

Solution

$$\overline{(\delta v)^2} = \frac{10^{12}\ \ 100^2 \cdot (10^{-6})^2 \quad 2 \cdot 10^{-4} \quad \Omega A^2}{10^{18}}$$

$$= 2 \times 10^{12+4-12-4-18}\ v^2$$

$$\sqrt{\overline{(\delta v)^2}} = 1.4 \times 10^9\ v$$

15 Light Sources

1. *Modes equation.* Derive equation (15.8),

$$k^2 = \pi^2\left[\left(\frac{M_X}{\ell_X}\right)^2 + \left(\frac{M_Y}{\ell_Y}\right)^2 + \left(\frac{M_Z}{\ell_Z}\right)^2\right] = \frac{\omega^2}{c^2}$$

similar to the two-dimensional derivation leading to equation 15.7.

Solution

We consider traveling waves in X, Y and Z direction

$$Ae^{i(k_X X \pm \omega t)}, \quad Ae^{i(k_Y Y \pm \omega t)}, \quad Ae^{i(k_Z Z \pm \omega t)}$$

The standing wave conditions in each direction are separately

$$2k_X\ell_X = M_X 2\pi, \quad 2k_Y\ell_Y = M_Y 2\pi, \quad 2k_Z\ell_Z = M_Z 2\pi$$

k_X, k_Y and k_Z are related to k using the angles θ and ϕ

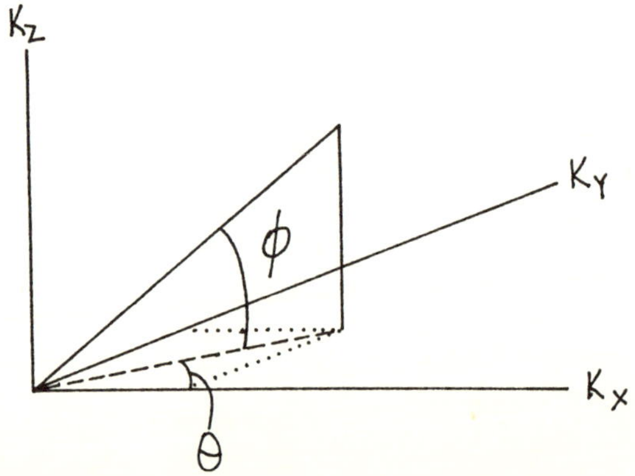

$k_X = k \cos\theta \cos\phi$ $\quad$ $k_Y = k \sin\theta \cos\phi$

$k_Z = k \sin\phi$

We have $k_X^2 + k_Y^2 + k_Z^2$

$$= k^2 (\cos^2\theta + \sin^2\theta)\cos^2\phi + k^2\sin^2\phi$$

$$= k^2$$

Introduction of the standing wave conditions yield

$$k^2 = \pi^2 \left[\left(\frac{M_X}{\ell_X}\right)^2 + \left(\frac{M_Y}{\ell_Y}\right)^2 + \left(\frac{M_Z}{\ell_Z}\right)^2\right]$$

2. *Radiation density equation.* Derive equation (15.12),

$$dZ = \tfrac{1}{8}(4\pi)R^2\,dR = 4\pi\frac{\ell^3}{c^3}\nu^2\,d\nu$$

similar to the two-dimensional case leading to equation 15.11.

15.2 Solution

We have $R = \sqrt{M_X^{\,2} + M_Y^{\,2} + M_Z^{\,2}} = \dfrac{2\ell\nu}{c}$

The number of combinations in the shell are

$$4\pi\ R^2\ dR$$

We have to use the factor $\frac{1}{8}$ and have

$$dZ = \frac{1}{8}\ 4\pi\ R^2\ dR = \frac{1}{8}\ 4\pi\ \frac{4\ell^2\nu^2}{c^2}\ \ \frac{2\ell}{c}\ d\nu$$

$$= 4\pi\ \frac{\ell^3}{c^3}\ \ \nu^2\ d\nu$$

3. *Elimination of ultraviolet catastrophe.* The Rayleigh-Jeans law diverges for small wavelengths. Consider Planck's formula,

$$\frac{du}{d\nu} = \frac{8\pi h\nu^3}{c^3}\frac{1}{e^{(h\nu/kT)} - 1}$$

Show that for Planck's formula, $\lim\limits_{\nu\to\infty}\dfrac{du}{d\nu}$ approaches 0, whereas in the Rayleigh-Jeans law it approaches ∞.

Solution

For $\nu\to\infty$ we may neglect 1 in comparison to $e^{\frac{h\nu}{kT}}$ and have

$$\lim_{\nu\to\infty}\ \frac{8\ \pi\ \ h}{c^3}\ \nu^3\ e^{-\frac{h\nu}{kT}}$$

using $x = \dfrac{h\nu}{kT}$ we have

$$\lim_{x\to\infty}\ \frac{8\ \pi\ \ h}{c^3}\ \ \left(\frac{kT}{h}\right)^3\ x^3\ e^{-x}$$

From mathematics we know that

$$\lim_{x\to\infty} x^n\ e^{-x} \to 0 \quad \text{for any finite number n}$$

Therefore $\lim_{\nu \to \infty} \left(\frac{du}{d\nu}\right) \to 0$

Rayleigh-Jeans:

$$\frac{du}{d\nu} = \frac{8 \pi k T}{c^3} \nu^2$$

$$\lim_{\nu \to \infty} \left(\frac{du}{d\nu}\right) \to \infty$$

4. *Rayleigh-Jeans law depending on wavelength.* The Rayleigh-Jeans law was derived as

$$\frac{du}{d\nu} = 8\pi k T \frac{\nu^2}{c^3}$$

Show that for the radiation density per wavelength interval we get

$$\frac{du}{d\lambda} = \frac{8\pi k T}{\lambda^4}$$

Solution

$\lambda \cdot \nu = c, \quad d\lambda \cdot \nu + d\nu \cdot \lambda = 0$

$$\left|\frac{d\nu}{d\lambda}\right| = \left|\frac{\nu}{\lambda}\right|$$

$$\frac{du}{d\nu} = 8\pi \; kT \frac{\nu^2}{c^3}$$

$$\left(\frac{du}{d\nu}\right) \cdot \left(\frac{d\nu}{d\lambda}\right) = 8\pi \; kT \frac{\left(\frac{c}{\lambda}\right)^2 \nu}{c^3 \; \lambda}$$

$$= \frac{8\pi \; kT \; c}{c \; \lambda^2 \lambda\lambda} = \frac{8\pi \; kT}{\lambda^4}$$

5. *Stefan-Boltzmann law.*
   a. Integrate $du/d\nu$ for Planck's law from 0 to ∞, and use the integral formula

$$\int_0^\infty \frac{x^3 \, dx}{e^x - 1} = \frac{\pi^4}{15}$$

With

$$\frac{W}{da} = \frac{c}{4} u$$

and show that for the radiant exitance we have

$$\frac{W}{da} = \frac{2}{15}\left(\frac{\pi^5 k^4 T^4}{c^2 h^3}\right)$$

b. Calculate the constant

$$\sigma = \frac{2}{15}\left(\frac{\pi^5 k^4}{c^2 h^3}\right)$$

to be $5.66 \times 10^{-8} (W/m^2 \cdot K^4)$.

c. Show that

$$\int_{\substack{\text{over}\\\text{sphere}}} \frac{da \cos\theta}{r^2} = 4\pi$$

Solution

(a) $$\frac{du}{d\nu} = \frac{8\pi h \nu^3}{c^3} \frac{1}{e^{\frac{h\nu}{kT}} - 1}$$

$$u = \int_0^\infty \frac{8\pi h\nu^3}{c^3} \frac{1}{e^{\frac{h\nu}{kT}} - 1} d\nu$$

We use $\nu = \frac{xkT}{h}$, $d\nu = \frac{kT}{h} dx$

$$u = \int_0^\infty \frac{8\pi h}{c^3} \left(\frac{kT}{h}\right)^3 \frac{x^3}{e^x - 1} \left(\frac{kT}{h}\right) dx$$

$$= \frac{8\pi h}{c^3} \left(\frac{kT}{h}\right)^4 \int_0^\infty \frac{x^3 dx}{e^x - 1}$$

$$u = \frac{8\pi h (kT)^4}{(ch)^3} \frac{\pi^4}{15} = \frac{8}{15} \frac{\pi^5 k^4 T^4}{c^3 h^3}$$

$$\frac{W}{da} = \frac{c}{4} = \frac{2}{15} \frac{\pi^5 k^4 T^4}{c^2 h^3}$$

(b) $$\sigma = \frac{2\pi^5 k^4}{15 c^2 h^3}$$

$k = 1.38 \times 10^{-23}$ J/K

$c = 2.998 \times 10^8$ m/sec

$h = 6.626 \times 10^{-34}$ J.sec

$$\sigma = \frac{2}{15} \frac{\pi^5 (1.38 \times 10^{-23} \text{ J/K})^4}{(2.998 \times 10^8 \text{ m/sec})^2 (6.626 \times 10^{-34} \text{ J sec})^3} =$$

$$5.66 \times 10^{-8} \frac{W}{m^2 K^4}$$

(c) We look from the volume element in the cavity to the cavity wall and have

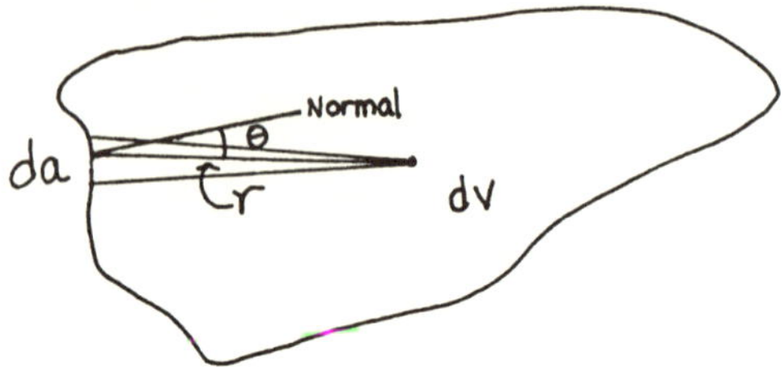

The solid angle $d\Omega$ from dV is $\dfrac{da \cos\theta}{r^2}$

and the integral over the surface S is the integral over the total spherical surface,

$$\int_{\text{over sphere}} d\Omega = 4\pi$$

6. *Wien's law.* Rewrite $du/d\nu$ as $du/d\lambda$, and show that for the maximum of $du/d\lambda$, we have

$$\lambda_{max} \cdot T = 2.899 \times 10^{-3}\,\mathrm{m \cdot K}$$

Solution

$$\frac{du}{d\nu} = \frac{8\pi h \nu^3}{c^3} \cdot \frac{1}{e^{\frac{h\nu}{kT}} - 1}$$

$$d\nu = d\left(\frac{c}{\lambda}\right) = \left|\frac{c}{\lambda^2}\right| d\lambda$$

$$\frac{du}{d\lambda} = \frac{8\pi h}{\lambda^3} \frac{1}{e^{\frac{hc}{kT\lambda}} - 1} \left(\frac{c}{\lambda^2}\right) = \frac{8\pi h c}{\lambda^5} \frac{1}{e^{\frac{hc}{kT\lambda}} - 1}$$

$$\frac{d}{d\lambda}\left(\frac{du}{d\lambda}\right) = 0 = 8\pi h c \left\{\frac{-5}{\lambda^6} \frac{1}{e^{\frac{hc}{kT\lambda}} - 1} + \frac{1}{\lambda^5}\left(\frac{(-1)\left(-\frac{hc}{kT\lambda^2}\right) e^{\frac{hc}{kT\lambda}}}{\left[e^{\frac{hc}{kT\lambda}} - 1\right]^2}\right)\right\}$$

We set $\frac{hc}{kT\lambda} = x$ and have

$$- \frac{5}{\lambda^6} \quad \frac{1}{e^x - 1} + \frac{\frac{1}{\lambda^6}\, x\, e^x}{(e^x - 1)^2} = 0$$

or $$\frac{x\, e^x}{e^x - 1} = 5$$

One can show (using a simple electronic calculator) that $x = 4.965$

$$\lambda_{max} = \frac{hc}{kT\ (4.965)}$$

or

$$\lambda_{max} \cdot T = \frac{hc}{k\ (4.965)}$$

$h = 6.626 \times 10^{-34}$ J·sec

$c = 2.998 \times 10^8 \ \frac{m}{sec}$

$k = 1.38 \times 10^{-23} \ \frac{J}{K}$

$$\lambda_{max} \cdot T = 2.899 \times 10^{-3} \text{ m} \cdot \text{K}$$

7. *Hydrogen atom, Rydberg constant, and transition frequencies.*
   a. The transition frequencies are given as

$$\nu_{n_i, n_f} = \frac{2\pi^2 K^2 e^4 m}{h^3}\left(\frac{1}{n_i^2} - \frac{1}{n_f^2}\right)$$

Using the definition $\hbar = h/2\pi$, where h is Planck's constant, we get

$$\frac{\nu_{n_i, n_f}}{c} = \left(\frac{1}{\lambda}\right)_{n_i, n_f} = \frac{K^2 e^4 m}{4\pi c \hbar^3}\left(\frac{1}{n_i^2} - \frac{1}{n_f^2}\right)$$

where m is the electron mass and c is the speed of light. Show that for

$$R = \frac{K^2 e^4 m}{4\pi c \hbar^3}$$

we have the value close to 109,720 cm^{-1} as originally given by Rydberg. Take $K = 8.988 \times 10\ N \cdot m^2/C$.

   b. The energy differences for the hydrogen atom can also be expressed as

$$\Delta E = E_{n_f} - E_{n_i} = E_0\left(\frac{1}{n_i} - \frac{1}{n_f}\right)$$

where

$$E_0 = \frac{2\pi^2 m K^2 e^4}{h^2} = 13.6 \text{ eV}$$

First calculate a table; then plot ΔE as a function of n_i for the Balmer series $n_f = 2$ for $n_i = 2$ to 10. The value of ΔE for $n_i \to \infty$ is the ionization energy for the hydrogen atoms with the electron originally in the state E_2.

Solution

(a)

$$K = 8.988 \times 10^9 \ N\cdot m^2/C^2$$

$$c = 2.998 \times 10^8 \ m/sec$$

$$e = 1.60 \times 10^{-19} \ C$$

$$\hbar = 1.055 \times 10^{-34} \ J\cdot sec$$

$$m_e = 9.1095 \times 10^{-31} \ kg$$

$$\frac{K^2 e^4 m_e}{c\cdot 4\pi\hbar^3} = \frac{(8.988 \times 10^9 \ N\cdot m^2/C^2)^2\cdot(1.60 \times 10^{-19} C)^4 (9.1095 \times 10^{-31} kg)}{(2.998 \times 10^8)\frac{m}{sec}\ (4)(1.055 \times 10^{-34} \ J\cdot sec)^3\cdot\pi}$$

$$= 1.090 \times 10^7 \ m^{-1}$$

(b)

$$\Delta E = E_o \left(\frac{1}{n_i^2} - \frac{1}{n_f^2}\right) \qquad E_o = 13.6 \ eV$$

$$n_f = 2$$

n_i	ΔE (in eV)
2	0
3	-1.89
4	-2.55
5	-2.85
6	-3.02
7	-3.12
8	-3.18
9	-3.23
10	-3.26

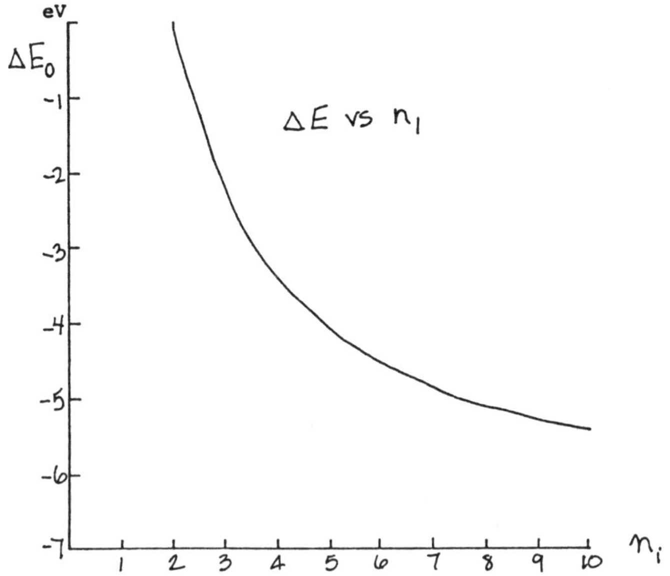

8. *Lorentz profile.* Calculate the width at half-height of an oscillator line (Lorentz). The intensity as a function of frequency is given as

$$I(\omega) = \frac{E_0^2}{4\pi^2((\omega - \omega_0)^2 + (1/\tau)^2)}$$

For the calculation, start from

$$\tfrac{1}{2}I(\omega_0) = I(\omega_{1/2})$$

where $\omega_{1/2}$ is the angular frequency for which the intensity has half its maximum.

Solution

$$\frac{1}{2} I(\omega_o) = \frac{\frac{1}{2} E_o^{\,2}}{4\pi^2 \left(\frac{1}{\tau^2}\right)} = \frac{E_o^{\,2}}{4\pi^2\left((\omega_{\frac{1}{2}} - \omega)^2 + \frac{1}{\tau^2}\right)}$$

$$\frac{2}{\tau^2} = (\omega_{\frac{1}{2}} - \omega)^2 + \frac{1}{\tau^2}$$

$$\frac{1}{\tau^2} = (\omega_{1/2} - \omega)^2 = \Delta\omega^2$$

$$\Delta\omega = \frac{1}{\tau}$$

9. *Doppler broadening.* Calculate the half-width for Doppler broadening. The intensity $I(\nu)$ is given as

$$I_{\nu_0} \exp^{[-1/2(mc^2/kT)(\nu-\nu_0/\nu_0)^2]}$$

as in problem 8, start with

$$\tfrac{1}{2} I(\nu_0) = I(\nu_{1/2})$$

<u>Solution</u>

$$\frac{1}{2} I(\nu_0) = \frac{1}{2} I_{\nu_0} = I_{\nu_0} \; e^{-\frac{1}{2}\frac{mc^2}{kT}\left(\frac{\nu_{1/2} - \nu_0}{\nu_0}\right)^2}$$

$$\ln \frac{1}{2} = -\frac{1}{2}\frac{mc^2}{kT}\left(\frac{\nu_{1/2} - \nu_0}{\nu_0}\right)^2$$

$$\ln 2 = \frac{1}{2}\frac{mc^2}{kT}\frac{1}{\nu_0^2}(\nu_{1/2} - \nu_0)^2$$

$$(\nu_{1/2} - \nu_0) = \nu_0 \sqrt{(\ln 2)\ \frac{2kT}{mc^2}}$$

$$\Delta\nu = 2(\nu_{1/2} - \nu_0) = 2\nu_0 \sqrt{\frac{2kT}{mc^2}\ln 2}$$

10. *Rate equation for three-level lasers*

a. Consider the steady state

$$dN_3/dt = 0 \quad \text{and} \quad dN_2/dt = 0 \text{ (for equation 15.67)}$$

derive

$$\frac{N_2}{N_1} = \left(\frac{W_{13}S_{32}}{W_{31} + S_{32}} + W_{12}\right)(A_{21} + W_{21})^{-1}$$

Note: A_{31} is small compared to W_{31}.

b. We know that $W_{12} = W_{21}$, $W_{31} = W_{13}$, and $S_{32} \gg W_{13}$. Derive

$$\frac{N_2}{N_1} = \frac{W_{13} + W_{12}}{W_{12} + A_{21}}$$

c. Consider

$$\frac{N_2 - N_1}{N_1} = \frac{W_{13} + W_{12}}{A_{21} + W_{12}} - 1$$

Use

$$\frac{N_0}{N_1} = \frac{N_1 + N_2 + N_3}{N_1}$$

and derive

$$\frac{N_2 - N_1}{N_0} = \frac{W_{13} - A_{21}}{W_{13} + A_{21} + 2W_{12}}$$

Solution

(a)

From $0 = W_{13} N_1 - (W_{31} + A_{31} + S_{32})N_3$

$$0 = W_{12} N_1 + S_{32}N_3 - (W_{21} + A_{21})N_2$$

we get

$$W_{13} N_1 = (W_{31} + A_{31} + S_{32})N_3$$

$$N_2(W_{21} + A_{21}) = W_{12} N_1 + S_{32} N_3$$

$$N_2 = \frac{W_{12} N_1 + S_{32}N_3}{W_{21} + A_{21}} = \frac{W_{12}}{W_{21} + A_{21}} N_1 + \frac{S_{32} W_{13}}{(W_{21} + A_{21})(W_{31} + A_{31} + S_{32})} N_1$$

$$\frac{N_2}{N_1} = \frac{W_{12}}{W_{21} + A_{21}} + \frac{S_{32} W_{13}}{(W_{21} + A_{21})(W_{31} + A_{31} + S_{32})}$$

$$= (W_{21} + A_{21})^{-1} \left\{\frac{W_{13} S_{32}}{W_{31} + A_{31} + S_{32}} + W_{12}\right\}$$

Since $A_{31} << W_{31}$

$$\frac{N_2}{N_1} = \left(\frac{W_{13} S_{32}}{W_{31} + S_{32}} + W_{12}\right)(W_{21} + A_{21})^{-1}$$

(b) We obtain from the expression in A with the conditions noted in Problem B.

$$\frac{N_2}{N_1} = \left(\frac{W_{13}\ S_{32}}{W_{31} + S_{32}} + W_{12}\right)(A_{21} + W_{21})^{-1}$$

$$\to \left(\frac{W_{13}\ S_{32}}{S_{32}} + W_{12}\right)(A_{21} + W_{12})^{-1}$$

$$\frac{N_2}{N_1} = \frac{W_{13} + W_{12}}{A_{21} + W_{12}}$$

(c) $$\frac{N_2 - N_1}{N_1}\ \frac{N_1}{N_o} = \left(\frac{W_{13} + W_{12}}{A_{21} + W_{12}} - 1\right)\frac{N_1}{N_o}$$

$$\frac{N_2 - N_1}{N_o} = \left(\frac{W_{13} + W_{12} - A_{21} - W_{12}}{A_{21} + W_{12}}\right)\frac{N_1}{N_o}$$

$$= \left(\frac{W_{13} - A_{21}}{W_{12} + A_{21}}\right)\frac{N_1}{N_o}$$

$$\frac{N_o}{N_1} = \frac{N_1}{N_1} + \frac{N_2}{N_1} + \frac{N_3}{N_1}$$

$$= 1 + \frac{W_{13} + W_{12}}{W_{12} + A_{21}} + \frac{W_{13}}{W_{13} + A_{31} + S_{32}}$$

since S_{32} is much larger than W_{13} , A_{31}

$$= 1 + \frac{W_{13} + W_{12}}{W_{12} + A_{21}} + \frac{W_{13}}{S_{32}}$$

the last term is small

$$\frac{N_o}{N_1} = 1 + \frac{W_{13} + W_{12}}{W_{12} + A_{21}}$$

$$\frac{N_2 - N_1}{N_o} = \left(\frac{W_{13} - A_{21}}{W_{12} - A_{21}}\right) \frac{1}{1 + \frac{W_{13} + W_{12}}{W_{12} + A_{21}}}$$

$$= \frac{(W_{13} - A_{21})(W_{12} + A_{21})}{(W_{12} + A_{21})(W_{12} + A_{21} + W_{13} + W_{12})}$$

$$\frac{N_2 - N_1}{N_o} = \frac{W_{13} - A_{21}}{W_{13} + A_{21} + 2W_{12}}$$

11. *Minimum power for laser oscillation.* Calculate the minimum power

$$NA_{21}h\nu[\mathrm{W \cdot cm^{-3}}]$$

to obtain laser oscillation for the unfavorable case where $A_{21} = \frac{1}{3} \times 10^8$, corresponding to equation 15.71. Assume that $h = 7 \times 10^{-34}$ W·s², $c = 3 \times 10^{10}$ cm/s, $N = 10^{19}$ cm^{-3}, and $\lambda = 0.5 \times 10^{-4}$ cm.

Solution

$$N\, A_{21}\, h\nu = \frac{10^{19}\mathrm{cm}^{-3}\ \frac{1}{3}\times 10^{8}\ \frac{1}{\mathrm{sec}}\ 7 \times 10^{-34}\ \mathrm{W\ sec}^2\ 3 \times 10^{10}\ \frac{\mathrm{cm}}{\mathrm{sec}}}{0.5 \times 10^{-4}\ \mathrm{cm}}$$

$$= \frac{10^{19+8-34+10+4} \times 7 \times 3}{3 \times 0.5} \quad \frac{\mathrm{watt}}{\mathrm{cm}^3}$$

$$= 14 \times 10^7 \cdot \frac{\mathrm{watt}}{\mathrm{cm}^3}$$

12. *Pulsed laser.* Consider a pulsed laser having pulse durations of 10^{-4} s and a peak power of 1 kW. If the repetition rate is 100 pulses per second, what is the comparable power of a continuous-wave laser?

Solution

$$\text{energy} = 1000\ \text{watt} \cdot 10^{-4}\ \text{sec} = \frac{1}{10}\ \text{watt sec}$$

$$\text{energy per 1 sec} = \frac{1}{10}\ \text{watt sec} \cdot 100\ \frac{1}{\text{sec}} = 10\ \text{watt}$$

16 Scattering of Light

1. *Randomly oriented particles.* For a collection of randomly oriented particles, compare the intensity of light scattering into the direction $X-Y$ ($\theta = 45°$) and Y ($\theta = 90°$) to what is scattered into the X ($\theta = 0°$) direction.

Solution

$$I(\theta) = \frac{1}{2} I_o(1 + \cos^2\theta)$$

$$I(0) = I_o$$

$$I(45) = \frac{1}{2} I_o(1 + \cos^2 45) = \frac{1}{2} I_o\ (1.5) = 0.75\ I_o$$

$$I(90) = \frac{1}{2} I_o$$

X - Y to X ratio 0.75

Y to X ratio 0.5

2. *Calculation of K.* Calculate equation 16.22 from equations 16.14, 16.19, and 16.21 by using

$$K = \frac{\Delta S}{S_F}\left(\frac{1}{\Delta x}\right) \quad \text{and} \quad v = F\,\Delta x$$

Solution

$$K = \frac{N\ \frac{4}{3}\ \mu_o\ \pi^3\ c^3\ p_o^{\ 2}}{\frac{1}{2}\ \varepsilon_o\ c\ E^2 F \cdot b\ \lambda^4\ \Delta x}$$

$$= \frac{N\ \frac{8}{3}\ \mu_o \pi^3\ c^2\ \left(\frac{e^2}{m\ \omega_o^{\ 2}}\right)^2\ E^2}{\varepsilon_o\ E^2 F\ b\ \lambda^4\ \Delta x}$$

$$= \frac{8}{3}\ \frac{N\ \mu_o}{\varepsilon_o}\ \frac{\pi^3\ c^2\ e^4}{m^2\ 16\pi^4\ \nu_o\ F\ b\ \lambda^4\ \Delta x}$$

$$= \frac{8}{3}\ \frac{N}{V}\ \frac{1 \cdot c^2\ e^4}{\varepsilon_o^{\ 2}\ c^2\ m^2\ 16\pi\ \nu_o^{\ 4}\ b\ \lambda^4}$$

$$= \frac{8}{3} N_v \frac{e^4}{\varepsilon_o^2} \frac{1}{m^2\, 16\pi\, \nu_o^4\, b\, \lambda^4}$$

$$= \left(\frac{N_o}{6} \frac{e^4}{\varepsilon_o^2} \frac{1}{m^2\, \pi\, \nu_o^4\, b}\right) \frac{1}{\lambda^4}$$

3. *Average of* $\cos^2 \theta$. For the dependence on angle θ for dipole emission, we have

$$|\mathbf{S}| \propto \cos^2 \theta$$

where θ is given as

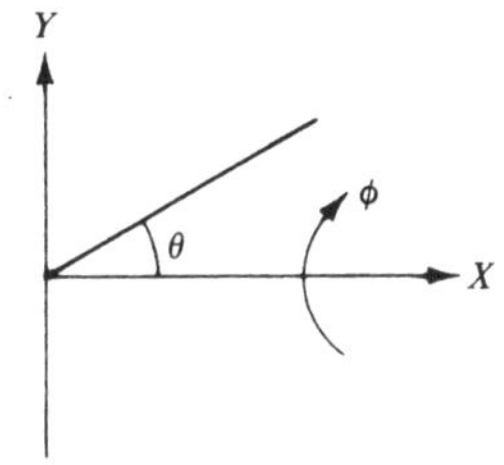

The angle ϕ is defined around the X axis. Calculate the average of $\cos^2 \theta$ over all directions in three dimensions.

<u>Solution</u>

$$Av = \int_0^{2\pi} \int_0^{\pi} \cos^2\theta \,\sin\theta d\theta d\phi$$

$$= -\,2\pi \int_0^{\pi} \cos^2\theta \; d(\cos\theta)$$

$$= -\,2\pi \left.\frac{\cos^3\theta}{3}\right|_0^{\pi} = 2\pi \left\{-\left(\frac{-1}{3} - \frac{1}{3}\right)\right\} = \frac{4\pi}{3}$$

4. *Bragg condition.* Consider a cubic crystal. The incident radiation has wavelength $\lambda = 1.6$ Å. The lattice constant is $a = 4.2$ Å. At what angle will we observe diffraction in first order ($n = 1$)?

<u>Solution</u>

$$\sin\theta = \frac{\lambda}{2a} = \frac{1.6}{8.4}$$

$$\theta = 10.98°$$

5. *Compton scattering.* For the scattering of an X ray by an electron, the energy of the photon is $E = h\nu$, the momentum of the photon is $p = h/\lambda$, the energy of the electron before scattering is $E_0 = mc^2$, where m is the rest mass of the electron, and the energy of the electron after scattering is $E = \sqrt{E_0^2 + p_e^2c^2}$, where c is the speed of light and p_e is the momentum of the electron after collision. We do not have to specify p_e because it will be eliminated in the derivation.

a. *Conservation of momentum.* We have the vector equation

$$\mathbf{p}_1 + \mathbf{p}_e = \mathbf{p}_2 + \mathbf{p}_e$$

$$(\text{before} = \text{after})$$

$$\mathbf{p}_e(\text{before}) = 0$$

and subscript 1 and 2 refer to "before" and "after" for the X ray, respectively. Show that we have

$$p_e^2 = p_1^2 + p_2^2 - 2p_1p_2\cos\theta$$

where θ is the angle between $\mathbf{p}_1$ and $\mathbf{p}_2$.

b. *Conservation of energy.*

$$p_1c + E_0 = p_2c + \sqrt{E_0^2 + p_e^2c^2}$$

Eliminate p_e^2 and obtain

$$\lambda_2 - \lambda_1 = \frac{h}{mc}(1 - \cos\theta)$$

Solution

(a)

$$\vec{p}_e^{\,2} = (\vec{p}_1 - \vec{p}_2)^2$$

$$= p_1^2 + p_2^2 - 2p_1 p_2 \cos\theta$$

(b)

$$((p_1-p_2)c + E_o)^2 = E_o^2 + c^2(p_1^2 + p_2^2 - 2p_1p_2 \cos\theta)$$

$$(p_1 - p_2)^2c^2 + 2(p_1 - p_2)c\, E_o + E_o^2 =$$

$$c^2p_1^2 - 2p_1p_2c^2 + p_2^2c^2 + 2(p_1 - p_2)c\, E_o + E_o^2$$

$$= E_o^2 + p_1^2c^2 + p_2^2c^2 - 2p_1p_2\, c^2 \cos\theta$$

$$E_o(p_1 - p_2) = p_1p_2\,(1 - \cos\theta)c$$

$$\frac{1}{p_2} - \frac{1}{p_1} = \frac{1}{mc^2}(1 - \cos\theta)c$$

$$\lambda_2 - \lambda_1 = \frac{h}{mc}(1 - \cos\theta)$$

6. *Linear chain.* Consider a free molecule having the masses

$$m_+ = m_H = 1.7 \times 10^{-27}\text{ kg} \quad \text{and} \quad m_- = m_{Li} = 11.4 \times 10^{-27}\text{ kg}$$

Assume that the force constant is $f = 7 \times 10^{-5}$ dyne/cm, corresponding to 7 N/m.

a. Calculate the vibrational frequency of this hypothetical molecule with

$$\nu = \frac{1}{2\pi}\sqrt{2f\left(\frac{1}{m_H} + \frac{1}{m_{Li}}\right)}$$

and give ν in cm^{-1} and the corresponding value λ_R in micrometers.

b. Using the same parameters, calculate the frequencies in cm^{-1} for $k = \pi/2a$, that is,

$$\nu_- = \frac{1}{2\pi c}\sqrt{\frac{2f}{m_-}} \quad \text{and} \quad \nu_+ = \frac{1}{2\pi c}\sqrt{\frac{2f}{m_+}}$$

Check the result for ν_+ by multiplying $\sqrt{m_-/m_+}$ by ν_-.

Solution

(a)

$$m_H = 1.7\text{x}10^{-27}\text{ kg}$$

$$m_{Li} = 11.4\text{x}10^{-27}\text{ kg}$$

$$f = 7\ \frac{N}{m}$$

$$\nu = \frac{1}{2\pi}\sqrt{2\cdot 7\ \frac{kg\ \frac{m}{sec^2}}{m}\ \left(\frac{1}{1.7} + \frac{1}{11.4}\right)\frac{1}{kg}\cdot 10\ \ (10^{13})}$$

$$= \frac{1}{2\pi}\sqrt{(14\cdot(0.68)\cdot 10\ 10^{13}}\ \frac{1}{sec}$$

$$\nu = 1.55\text{ x }10^{13}\ \frac{1}{sec}$$

$$\nu_{cm^{-1}} \Rightarrow \frac{\nu}{c} = \frac{(1.55)\text{x}10^{13}}{3\text{x}10^{10}}\ \frac{1}{cm} = 517\ cm^{-1}$$

$$\lambda_{in\ \mu} = 19.34\mu$$

(The value for the solid is in this wavelength range at around 13μ.)

(b)

$$\nu_- = \frac{1}{2\pi c}\ \sqrt{\frac{2f}{m_-}} = \frac{1}{2\cdot 3\text{x}10^{10}}\ \sqrt{\frac{2\cdot 7}{11.4\text{x}10^{-27}}}$$

$$= \frac{\sqrt{12.3 \times 10^{13}}}{2\pi \cdot 3 \times 10^{10}} = \frac{3.5}{2 \cdot (3.14) \cdot 3} 10^{3} = 186 \text{ cm}^{-1}$$

$$\nu_+ = \frac{1}{2\pi c} \sqrt{\frac{2 \cdot 7}{1.7} 10} \cdot 10^{13} = 481 \text{ cm}^{-1}$$

$$\nu_- \sqrt{\frac{m_-}{m_+}} = \nu_+ = 186 \cdot (2.59) = 481 \text{ cm}^{-1}$$

7. *Brillouin scattering.*
   a. Calculate the shifted frequency ω' from $\omega - \omega' = \Omega$. First calculate

$$\Omega = 2v_s \frac{2\pi}{\lambda} n \sin\frac{\phi}{2}$$

   Using $v_s = 5.5 \times 10^5$ cm/s, $\lambda = 4358.3$ Å (mercury lamp line), $n = 1.55$, and $\sin(\phi/2) = 1$.
   b. Calculate $\omega = 2\pi\nu$ from $\lambda\nu = c$ for $\lambda = 4358.3$ Å; use $c = 2.9979 \times 10^{10}$ cm/s.
   c. Calculate $\omega' = \omega - \Omega$.
   d. Convert Ω into λ_s (in angstroms) and ω' into λ' (in angstroms) and compare the difference corresponding to $\omega - \omega'$ to λ_s.

Solution

(a) $$\Omega = 2 \times (5.5) \times 10^{5} \frac{\text{cm}}{\text{sec}} \frac{2 \cdot (3.14)}{0.43583 \times 10^{-4} \text{ cm}} 1.55$$

$$= 245.8 \times 10^{9} \frac{1}{\text{sec}} = 2.46 \times 10^{11} \frac{1}{\text{sec}}$$

(b) $$\omega = 2\pi\nu = 2\pi \frac{c}{\lambda} = 2\pi \frac{2.9979 \times 10^{10} \frac{\text{cm}}{\text{s}}}{.43583 \times 10^{-4} \text{ cm}}$$

$$= 2\pi \; 6.8786 \times 10^{14} \frac{1}{\text{sec}}$$

$$\omega = 43.2195 \times 10^{14} \frac{1}{\text{sec}}$$

(c)

$$\omega' \Rightarrow \omega = 43219.5 \times 10^{11} \ \frac{1}{sec}$$

$$-\Omega = - \quad 2.5 \times 10^{11} \ \frac{1}{sec}$$

$$\omega' = 43217.0 \times 10^{11} \ \frac{1}{sec}$$

$$\omega' = 43.2170 \times 10^{14} \ \frac{1}{sec}$$

(d)

$$\Omega \lambda_s = 2\pi v_s \qquad \lambda_s = \frac{2\pi \cdot (5.5) \times 10^5 \ \frac{cm}{sec}}{(2.46) \times 10^{11} \ \frac{1}{sec}}$$

$$\lambda_s = .14 \times 10^{-8} \ cm$$

$$\lambda' = \frac{2\pi c}{\omega'} = \frac{2\pi \cdot (2.9979) \times 10^{10} \ \frac{cm}{sec}}{(43.2170) \times 10^{14} \ \frac{1}{sec}}$$

$$= 0.43586 \times 10^{-4} \ cm$$

$$\omega \rightarrow \lambda = 4358.3 \ \text{Å}$$

$$\omega' \rightarrow \lambda' = 4358.6 \ \text{Å}$$

$$\Omega \rightarrow \lambda_s = .14 \ \text{Å}$$

8. *Raman effect.* The exciting line for the Raman effect is the 4358.3 Å line of mercury. The vibrational frequencies of CCl_4 are $v_2 = 218 \text{ cm}^{-1}$, $v_4 = 314 \text{ cm}^{-1}$, $v_1 = 458 \text{ cm}^{-1}$, and $v_3 = 775 \text{ cm}^{-1}$. Assume that we want to give an interpretation of a spectrograph in an angstrom scale. Calculate the four Raman lines for CCl_4 in angstroms.

Solution

The line 4358.3 Å corresponds to 22944.7 cm^{-1}.

The Stokes lines are at lower frequencies, we have

$$\nu_2 = 22726 \text{ cm}^{-1}$$

$$\nu_4 = 22629 \text{ cm}^{-1}$$

$$\nu_1 = 22486 \text{ cm}^{-1}$$

$$\nu_3 = 22169 \text{ cm}^{-1}$$

The corresponding wavelengths are

$$\lambda_2 = 4400$$

$$\lambda_4 = 4419$$

$$\lambda_1 = 4447$$

$$\lambda_3 = 4511$$

---

17 Integrated Optics

1. *Angle of k vector.* The angle between the k vector (direction of propagation of the wave) and the normal of the plates of a cavity was called θ. We may approximate

$$\tan\theta = \frac{\overline{m_X}/\ell}{\overline{m_Y}/d}$$

Calculate the angle θ for

$$\overline{m_X} = 50; \qquad \overline{m_Y} = 1$$
$$\overline{m_X} = 50; \qquad \overline{m_Y} = 2$$
$$\overline{m_X} = 25; \qquad \overline{m_Y} = 1$$
$$\overline{m_X} = 25; \qquad \overline{m_Y} = 2$$

Two of those modes have the same angle θ.

Solution

$$\arctan \frac{1}{100}\ \frac{50}{1} = 26.6°$$

$$\arctan \frac{1}{100}\ \frac{50}{2} = 14°$$

$$\arctan \frac{1}{100}\ \frac{25}{1} = 14°$$

$$\arctan \frac{1}{100}\ \frac{25}{2} = 7.1°$$

2. *Phase velocity*. For $\theta = 26°$, 14°, and 7°, calculate the phase velocity in a medium of refractive index $n_2 = 1.7$. Express v_X and v_Y as multiples of $c = 3 \times 10^8$ m/s in vacuum.

<u>Solution</u>

$$v_X = \frac{1}{1.7 \cdot \sin 26} c = 1.34\ c$$

$$v_X = \frac{1}{1.7 \cdot \sin 14} c = 2.43\ c$$

$$v_X = \frac{1}{1.7 \cdot \sin 7} c = 4.82\ c$$

$$v_Y = \frac{1}{1.7 \cdot \cos 26} c = 0.65\ c$$

$$v_Y = \frac{1}{1.7 \cdot \cos 14} c = 0.60\ c$$

$$v_Y = \frac{1}{1.7 \cdot \cos 7} c = 0.59\ c$$

3. *Tapered section of a thin-film waveguide.* Consider a tapered section having an angle α with the horizontal.

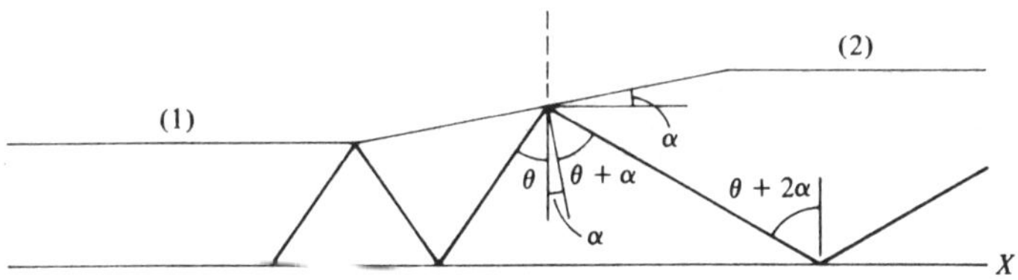

The angle θ is changing from θ to $\theta + 2\alpha$. How is the phase velocity v_X changing if $\alpha = 2°$ and $\theta = 14°$? Give the change as a multiple of $c = 3 \times 10^8$ m/s and observe that $v_1 - v_2 = \Delta v$ is negative, that is, the phase velocity decreases going from section 1 to section 2.

<u>Solution</u>

In 1 $\quad v_X = \dfrac{c}{1.7 \quad \sin 14} = 2.43\ c$

In 2 $\quad v_X = \dfrac{c}{1.7 \quad \sin 18} = 1.90\ c$

$$\Delta v = -\ 0.53\ c$$

4. *Conical light pipe.* Consider a conically shaped tube with a large opening of radius r_1, small opening of radius r_2, and length x. The inner surfaces are perfectly reflecting. Consider only rays in a plane containing the axis. We want to investigate the relationship between the angle at which an extreme ray enters the cone, α_1, and the angle at which it emerges, α_2. To determine α_2, we do not have to trace the ray through the cone, but can "fold" the cone and consider a straight-line path as shown in Figure b. If the ray is tangent to the dotted circle, $\alpha_2 = 90°$ and all rays with incident angles less than the corresponding α_1 will pass through the cone. If $\alpha_2 > 90°$, some rays will be reflected backward and not pass through.

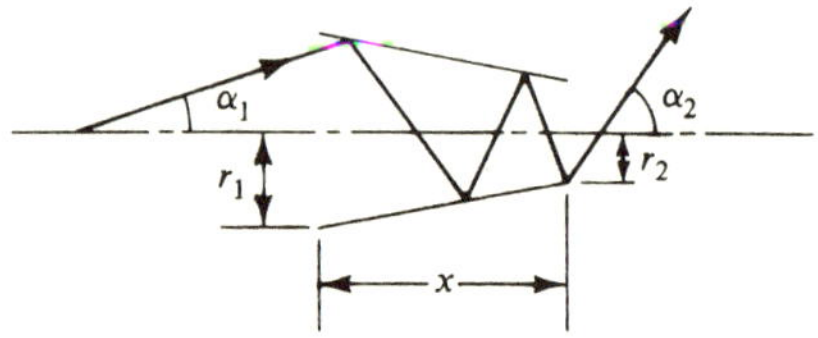

(a)

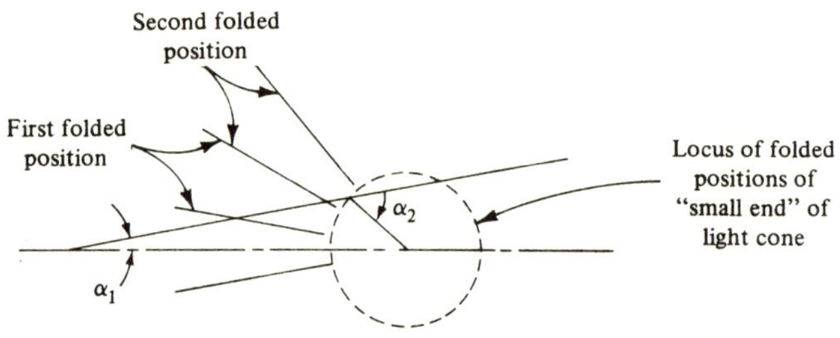

(b)

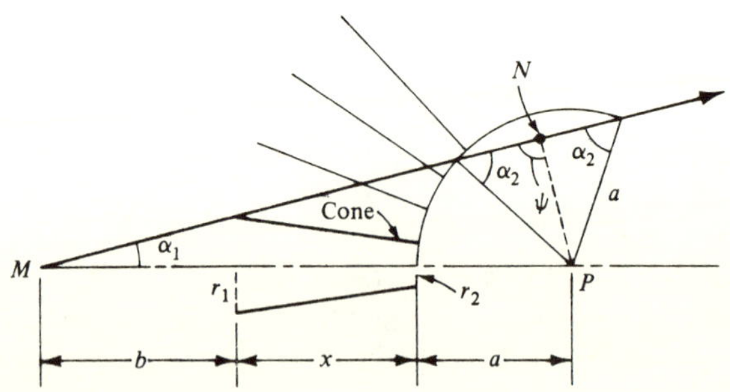

(c)

a. Derive the following expression relating x, r_1, r_2, α_1, and α_2 by referring to the figure above.

$$x = \left(1 - \frac{r_2}{r_1}\right) \frac{r_1 \cos \alpha_1}{(r_2/r_1) \sin \alpha_2 - \sin \alpha_1}$$

One may first deduce the following relations:

$$a = \frac{x}{(r_1/r_2) - 1} \qquad \tan \alpha_1 = \frac{r_1}{b} \qquad \frac{\sin \alpha_1}{a \sin \alpha_2} = \frac{1}{b + x + a}$$

b. Show that for the case where $x \to \infty$, we obtain

$$r_2 \sin \alpha_2 = r_1 \sin \alpha_1$$

This corresponds to Abbe's sine condition and tells us that for a long tube (e.g., optical fibers), the image quality is conserved by transmission.

Solution

(a) Similar Δ's => $\frac{r_1}{x + a} = \frac{r_2}{a}$ => $x + a = \frac{r_1}{r_2} a$

$$a(\frac{r_1}{r_2} - 1) = x \Rightarrow a \frac{x}{\frac{r_1}{r_2} - 1}$$

From the figure:

$$\tan \alpha_1 = \frac{r_1}{b}$$

$\psi = \pi/2$ by definition

From Δ MNP and the law of sines

$$\frac{\sin \alpha_1}{\overline{PN}} = \frac{\sin \psi}{b + x + a}$$

As $\psi = \pi/2$ one has $\overline{PN} = a \sin \alpha_2$;

then $$\frac{\sin \alpha_1}{a \sin \alpha_2} = \frac{1}{b + x + a}$$

we want to eliminate a and b:

$$\frac{\sin \alpha_1 \left(\frac{r_1}{r_2} - 1\right)}{x \sin \alpha_2} = \frac{1}{r_1/\tan \alpha_1) + x + x/\left(\frac{r_1}{r_2} - 1\right)}$$

$$\left(\frac{r_1}{r_2} - 1\right) \sin \alpha_1 \left[(r_1/\tan \alpha_1) + x + x/\left(\frac{r_1}{r_2} - 1\right)\right] = x \sin \alpha_2$$

$$x \left\{\left(\frac{r_1}{r_2} - 1\right) \sin \alpha_1 + \sin \alpha_1 - \sin \alpha_2\right\} = -\left(\frac{r_1}{r_2} - 1\right) \frac{\sin \alpha_1 \, r_1}{\tan \alpha_1}$$

$$x = \frac{-\left(\frac{r_1}{r_2} - 1\right) r_1 \cos \alpha_1}{\frac{r_1}{r_2} \sin \alpha_1 - \sin \alpha_2} = \frac{\left(\frac{r_1}{r_2} - 1\right) r_1 \cos \alpha_1}{\sin \alpha_2 - \frac{r_1}{r_2} \sin \alpha_1} =$$

$$\Rightarrow \quad x = \left[\frac{\left(1 - \frac{r_2}{r_1}\right) r_1 \cos \alpha_1}{\frac{r_2}{r_1} \sin \alpha_2 - \sin \alpha_1}\right]$$

(b) For $x \to \infty \to [\frac{r_2}{r_1} \sin \alpha_2 - \sin \alpha_1] \to 0$

$\Rightarrow r_2 \sin \alpha_2 = r_1 \sin \alpha_1$

For $n_2 = n_1 = 1$ i.e. air in the cone. This corresponds to Abbe's sine condition.

5. *Grating coupler.* Consider the diffraction on an amplitude grating as discussed in problem 6 of Chapter 3. There we found $a(\sin\theta + \sin\phi) = m\lambda$ for the condition of constructive interference.

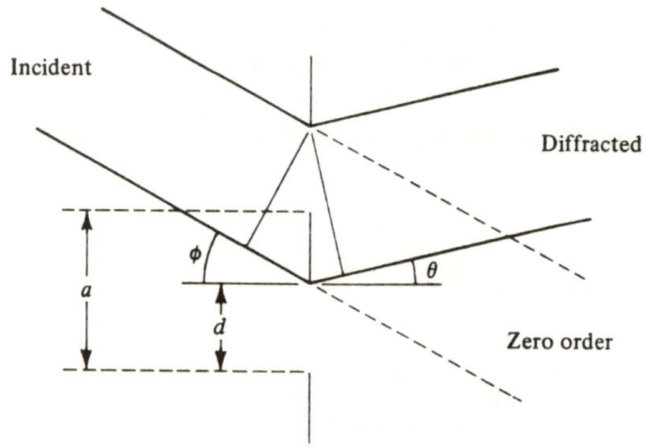

Now consider light from the left under angle $\phi = 0°$. The right side has the refractive index n_2.

a. Show that for the condition of constructive interference at normal incidence ($\phi = 0°$), we get

$$a(n_2 \sin\theta) = m\lambda$$

b. Assume now that the wave on the left is incident under the angle ϕ and that the refractive index on the left is n_1. Show that we now have

$$a(n_2 \sin\theta + n_1 \sin\phi) = m\lambda$$

We can obtain the k values for the different orders of diffraction by considering $k = (2\pi/\lambda)n$ and calculate λ from the above conditions.

c. Calculate k for $m = 0, 1, 2$.

<u>Solution</u>

(a)

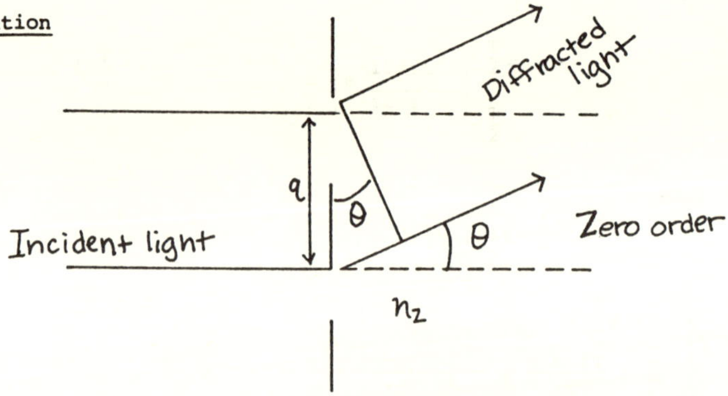

The path difference $a \sin\theta$ is now equal to $m \frac{\lambda}{n_2}$,

that is

$$a\, n_2 \sin\theta = m\lambda$$

(b)

The path difference on the left is now $a \sin\phi$ and has to be multiplied by m_1 to account for the optical path.

$$a\, n_2 \sin\theta + a\, n_1 \sin\phi = m\lambda$$

(c)

$$k = \frac{2\pi}{\lambda}\, n$$

$m = 0$

$$k_o = \frac{2\pi}{\lambda}\, n_2$$

direction by law of refraction

$$n_2 \sin\theta + n_1 \sin\phi = 0$$

$m = 1$ $\quad a(n_2 \sin\theta + n_1 \sin\phi) = \lambda$

$$k_1 = \frac{2\pi\, n_2}{a(n_2 \sin\theta + n_1 \sin\phi)}$$

$m = 2$ $\quad a(n_2 \sin\theta + n_1 \sin\phi) = 2$

$$k_2 = \frac{4\pi\, n_2}{a(n_2 \sin\theta + n_1 \sin\phi)}$$

6. *Decibal problem.* The attenuation is given in decibels, and we have $1\,\text{dB} = 10 \log_{10} S/N$.
   a. Give the ratio S/N for 20 dB.
   b. In optics, we present the attenuation in experiments as $I = I_0 e^{-\alpha d}$, where α is in 1/cm when the path length is given in centimeters. How many decibels correspond to the attenuation if $\alpha = 0.1\ \text{cm}^{-1}$ and $d = 10$ cm?

Solution

(a)

$$\log_{10} \frac{S}{N} = \frac{20}{10} = \quad \rightarrow \quad \begin{matrix} S = 100 \\ N = 1 \end{matrix}$$

(b)

$$I = I_o e^{-0.1 \cdot 10} = I_o e^{-1}$$

$$\frac{I_o}{I} = e \qquad \log_{10} e = \log_{10} 2.72 = 0.434$$

$$10 \log_{10} \frac{I_o}{I} = 4.34 \text{ db}$$

7. *Ray equation.* Write out the formulas for the derivation of equations 17.22 and 17.23.

Solution

$$(n \vec{s}_o) \cdot \frac{d}{ds} (\vec{s}_o n) = f(y) \; n(\vec{s}_o \text{ grad } n)$$

For the left side we may write

$$\frac{1}{2} \frac{d}{ds} (n s_o)^2 = \frac{1}{2} \frac{d}{ds} n^2 = n \frac{dn}{ds}$$

We have

$$n \frac{dn}{ds} = f(y) \; n(\vec{s}_o \text{ grad } n)$$

$$(\vec{s}_o \text{ grad } n) = \frac{dn}{ds}$$

and we have

$n = f(y)\, n$

$f(y)$ must be 1.

Therefore we get in $\vec{s}_o$ direction

$$n \frac{d}{ds} (\vec{s}_o\, n) = n \text{ grad } n$$

or

$$\frac{d}{ds} (\vec{s}_o\, n) = \text{grad } n$$

8. *Index of refraction at Y_{max} for a fiber.* Consider the index of refraction $n = n_0(1 - 2\eta^2 Y^2)$.
   a. Show that the value for Y_{max} depending on θ (angle with the normal) may be expressed as

$$Y_{max} = \frac{(\cot\theta)L}{2\pi}$$

   b. Show that the variation of the refraction index depending on Y_{max} i

$$n_{opt} = n_0(1 - \tfrac{1}{2}\cot^2\theta)$$

   c. Calculate the refractive index ratio n_{opt}/n_0 if we launch the ray with angles 2°, 4°, and 6° to the horizontal.

Solution

(a)

$$Y = A \sin 2\eta X$$

$$\left.\frac{dY}{dX}\right|_{X=0} = \cot\theta$$

$$\left.\frac{dY}{dX}\right|_{X=0} = A(\cos 2\eta X)\, 2\eta \Big|_{X=0}$$

$$A 2\eta = \cot\theta \qquad L\eta = \pi$$

$$Y_{max} = \frac{L \cot\theta}{2\pi}$$

(b)

$$n_{opt} = n_o \ (1 - 2\eta^2 \ Y_{max}^{\ 2})$$

$$= n_o \ (1 - 2\eta^2 \ \frac{\cot \ \theta^2 L^2}{4\pi^2}) \quad \text{with } L\eta = \pi$$

$$= n_o \ (1 - 2 \ \frac{\pi^2}{L^2} \quad \frac{(\cot \ \theta)^2 L^2}{4\pi^2})$$

$$= n_o \ (1 - \frac{1}{2} \cot^2 \ \theta)$$

(c)

We launch the ray under an angle of 2°, 4°, 6° with respect to the X axis, that corresponds to θ = 88°, 86°, 84°.

Since cot θ = tan (90 - θ) we have for

$$\frac{n}{n_o} = (1 - \tfrac{1}{2}(\tan 2°)^2) = 1 - 0.00061$$

$$\frac{n}{n_o} = (1 - \tfrac{1}{2}(\tan 4°)^2) = 1 - 0.00245$$

$$\frac{n}{n_o} = (1 - \tfrac{1}{2}(\tan 6°)^2) = 1 - 0.00552$$

18 Acoustooptics and Nonlinear Effects

1. *Continously changing refractive index.* A light ray enters a medium of refractive index n_Δ from a medium of refractive index n.

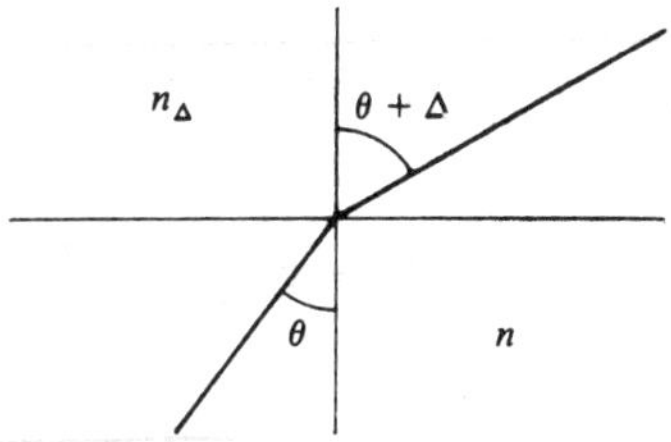

We want to calculate the refractive index n_Δ if θ is between 50° and 80° and the change in θ is $\Delta = 1°$. Calculate n_Δ/n for $\theta = 50°, 60°, 70°, 80°$.
(*Note:* You will find that n_Δ changes less for larger angles but more for smaller angles for 1° derivation. A continuously decreasing refractive index would bend the ray.)

Solution

$$50°: n_\Delta = .9857\ n$$

$$60°: n_\Delta = .9902\ n$$

$$70°: n_\Delta = .9938\ n$$

$$80°: n_\Delta = .9971\ n$$

2. *Scattering on an acoustic wave.*
   a. Check the calculation for $\omega_2 - \omega_1 = \Omega_s$ (equation 18.3).
   b. Check the orders of magnitude of ω and Ω_s for

$$\lambda = 0.5 \times 10^{-4}\ \text{cm}, \quad c = 3 \times 10^{10}\ \text{cm/s}$$

and

$$\lambda_s = 1.6 \times 10^{10}\ \text{cm}, \quad v_s = 5 \times 10^5\ \text{cm/s}$$

Solution

(a)

$$\omega_2 - \omega_1 = \frac{2\ \omega_2\ v_s\ n \sin\alpha}{c}$$

$$v_s = \lambda_s \frac{\Omega_s}{2\pi} \qquad \sin\alpha = \frac{\lambda_2}{2\ \lambda_s}$$

$$\omega_2 = 2\pi \frac{\frac{c}{n}}{\lambda_2}$$

$$\omega_2 - \omega_1 = \frac{2 \cdot 2\pi \frac{\frac{c}{n}}{\lambda_2} \lambda_s \frac{\Omega_s}{2\pi} n \frac{\lambda_2}{2\lambda_s}}{c} = \Omega_s$$

(b)

$$\lambda = .5 \cdot 10^{-4} \text{ cm} \rightarrow \omega = 38 \cdot 10^{14} \frac{1}{\text{sec}}$$

$$\lambda_s = 1.6 \cdot 10^{10} \text{ cm} \rightarrow \Omega = 2 \times 10^{-4} \frac{1}{\text{sec}}$$

3. *Bragg condition.*
   a. Go through the derivation of equation 18.10.
   b. Calculate α' from $\alpha = 25°$, $\nu_S = 10^{10}$ 1/s, $v_S = 5 \times 10^5$ m/s, $\nu = 3 \times 10^{15}$ 1/s, and $c = 3 \times 10^{10}$ m/s.
   c. As a check, calculate from $\lambda/\lambda_S = 0.2$ the value of λ_S for λ corresponding to $\nu = 3 \times 10^{15}$ 1/s, and (as a check) then calculate ν_S from λ_S.

Solution

(a) $$\frac{\lambda}{\lambda_S} = \sin\alpha + \sin\alpha'$$

$$\sin\alpha' = \frac{\lambda}{\lambda_S} - \sin\alpha$$

$$= \frac{\frac{c}{\nu}}{\frac{v_S}{\nu_S}} - \sin\alpha$$

$$\sin\alpha' = -\sin\alpha + \frac{c\nu_S}{v_S \nu}$$

(b)

$$\sin\alpha + \sin\alpha' = \frac{3\times10^{10} \frac{\text{cm}}{\text{sec}} \; 10^{10} \frac{1}{\text{sec}}}{5\times10^{5} \frac{\text{cm}}{\text{sec}} \; 3\times10^{15} \frac{1}{\text{sec}}}$$

$$= \frac{3}{15} 10^{10+10-15-5} = 0.2$$

$\sin 25° = 0.423$

$\sin \alpha' = 0.2 - 0.423 = -0.223$

$\alpha' = -12.8°$

(c)

$$\nu = 3\text{x}10^{15} \frac{1}{\text{sec}} \rightarrow \lambda = \frac{3\text{x}10^{10} \frac{\text{cm}}{\text{sec}}}{3\text{x}10^{15} \frac{1}{\text{sec}}}$$

$$\lambda_S = 5\text{x}10^{-5} \text{ cm}$$

$$\nu_S = \frac{v_S}{\lambda_S} = \frac{5\text{x}10^5 \frac{\text{cm}}{\text{sec}}}{5\text{x}10^{-5}\text{cm}} = 10^{10} \frac{1}{\text{sec}}$$

4. *Michelson echelon grating.* Consider the echelon grating used in the "forward" direction (i.e., $\theta = 0°$).

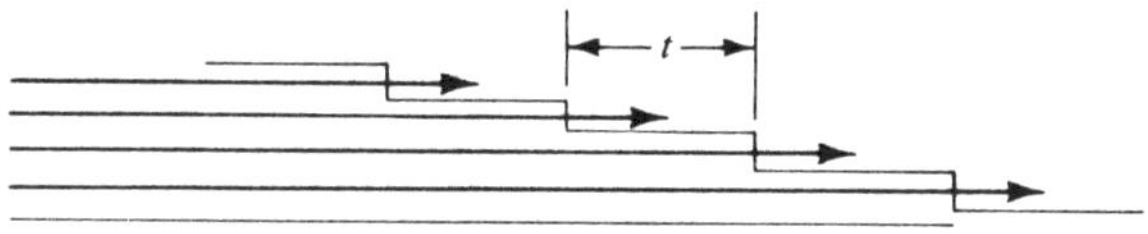

The glass has refractive index $n = 1.5$, and $t = 18$ mm. The light used has wavelength $\lambda = 0.5 \times 10^{-4}$ cm. Show that we have constructive interference for an order number of 18,000.

Solution

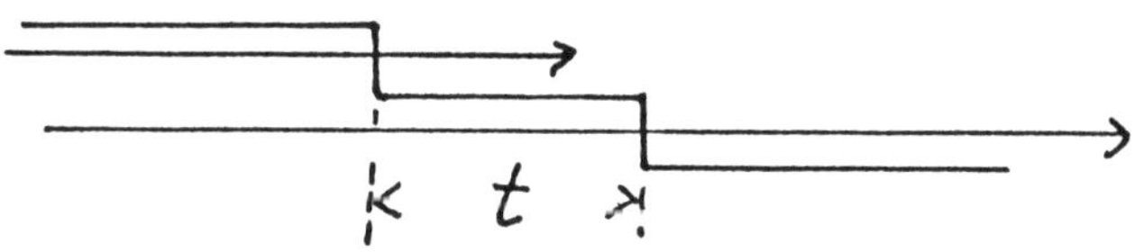

$n = 1.5$

path difference = $(n-1)t$

path difference between neighboring rays is equal $(n-1)t$

This must be equal to $m\ \lambda$. Therefore

$$m = \frac{(1.5-1)\ 18\text{mm}}{(0.5)\text{x}10^{-4}\ \text{cm}}$$

$$= \frac{(0.5)\ (1.8)\text{cm}}{(0.5)\text{x}\ 10^{-4}\text{cm}} = 1.8\text{x}10^4$$

$m = 18000$

5. *Raman effect and Stokes and anti-Stokes lines.* The Boltzmann distribution gives us the fraction of molecules in the state E at temperature T as

$$N = N_0 e^{-(E/kT)}$$

In the Raman effect, the molecules absorb the incident photon of energy $h\nu$. In the emission process, the energy of a molecular vibration $h\nu_v$ is either added or subtracted. If the emitted energy of the photon is lower, we have $E_S = h(\nu - \nu_v)$; if it is higher, we have $E_{AS} = h(\nu + \nu_v)$, where S stands for Stokes and AS for anti-Stokes emission. According to the Boltzmann distribution, we have

$$N_S > N_{AS}$$

Calculate N_S/N_{AS} using $\bar{\nu}_v = 500\ \text{cm}^{-1}$, $\bar{\nu} = 22938\ \text{cm}^{-1}$, $k = 1.4 \times 10^{-23}$ J/K, $T =$ 300 K, and $h = 6.6 \times 10^{-34}$ J · s, and show that for this case we have

$$\frac{N_S}{N_{AS}} \simeq 100$$

Note:

$$1\ \text{cm}^{-1} = \frac{\nu\ 1/\text{s}}{c\ \text{cm/s}}$$

where $c = 2.9979 \times 10^{10}$ cm/s.

Solution

(a) We convert ν and ν_V into frequencies by multiplication with $c = 3 \times 10^{10}\ \frac{\text{cm}}{\text{sec}}$

$$\nu_V = \bar{\nu}_V c = 500 \cdot 3 \times 10^{10}\ \frac{1}{\text{sec}}$$

$$\nu = \bar{\nu} c = 229 \cdot 3 \times 10^{10}\ \frac{1}{\text{sec}}$$

$$\nu_V = 1500 \times 10^{10}\ \frac{1}{\text{sec}}$$

$$\nu = 68814 \times 10^{10}\ \frac{1}{\text{sec}}$$

$$\nu_+ = \nu + \nu_V = 70314 \times 10^{10}\ \frac{1}{\text{sec}}$$

$$\nu_- = \nu - \nu_V = 67314 \times 10^{10}\ \frac{1}{\text{sec}}$$

$$\frac{h\nu_+}{kT} = \frac{(6.6) \times 10^{-34}\ \text{J sec} \cdot 70314 \times 10^{10}\ \frac{1}{\text{sec}}}{(1.4) \times 10^{-23}\ \frac{\text{J}}{\text{K}}\ 300\text{K}}$$

$$= \frac{(6.6) \cdot 70314}{(1.4) \cdot 300}\ 10^{-34+10+23}$$

$$= 1104 \times 10^{-1} - 110.4$$

$$\frac{h\nu_-}{kT} = \frac{(6.6) \times 10^{-34} \text{ J sec} \cdot 67314 \times 10^{10} \frac{1}{\text{sec}}}{(1.4) \times 10^{-23} \frac{J}{K} \cdot 300K}$$

$$= \frac{(6.6) \cdot 67314}{(1.4) \cdot 300} \quad 10^{-34+10+23}$$

$$\frac{h\nu_-}{kT} = 1057 \times 10^{-1} = 105.7$$

$$\frac{N_S}{N_{AS}} = e^{110.4 - 105.7} = e^{4.7} \approx 110$$

6. *Power for self-focusing.* Yariv gives the power level for self-focusing as

$$P_C = \frac{\varepsilon_0 n^3 c\lambda^2}{4\pi n_2'}$$

where the total dielectric constant is given as

$$\varepsilon_{\text{total}} = \varepsilon + \varepsilon_2 E^2$$

E being the electric field. The following values are given:

$$n = \sqrt{\frac{\varepsilon}{\varepsilon_0}}$$

$$(n_2')_{\text{esu}} = 10^{-11}, \quad \lambda = 10^{-6} \text{ m}$$

$$\varepsilon \simeq \varepsilon_0, \quad (n_2')_{\text{esu}} = 9 \times 10^8 (n_2')_{\text{mks}}$$

$$\varepsilon_0 = 8.25 \times 10^{-12} \frac{\text{C}^2}{\text{N} \cdot \text{m}^2}$$

Yariv arrives at $P_C \simeq 2 \times 10^4$ W. Show that the value is correct.

<u>Solution</u>

(a)

$$P_C = \frac{8.85 \times 10^{-12} \cdot 3 \times 10^8 \frac{m}{s} \; 10^{-12} \; m^2 \; C^2}{4 \cdot (3.14) \; (10^{-11} \frac{1}{9} 10^{-8}) \text{ mKs} \quad Nm^2}$$

$$= \frac{(8.85) \cdot 3 \cdot 9}{4 \cdot (3.14)} \quad 10^{-12+8-12+11+8} \quad W$$

$$= 1.9 \times 10^4 \text{ W}$$

7. *Nonlinear polarizability.* Consider the polarizability of a hydrogen-like atom. We have

$$P_a = \alpha E$$

For hydrogen, we have $\alpha = 1.7 \times 10^{-41}$ mC^2/N. We take $E = 5 \times 10^5$ N/C (mks units).

a. Calculate P_a.

We have expressed the polarizability in one dimension as

$$P = \alpha_1 E + \alpha_2 E^2$$

In three dimensions, P and E are vectors and α_2 are tensors. In crystals of point group symmetry $\bar{4}2$ m, we have for the three components of the nonlinear polarizability

$$P_X = 2d_{14}E_Z E_Y$$
$$P_Y - 2d_{14}E_Z E_X$$
$$P_Z = 2d_{36}E_X E_Y$$

where E_X, E_Y, and E_Z are the components of the electric field and d_{14} and d_{36} are constants. In particular, for KDP we have $d_{14} = d_{36}$. For the conversion, we have

$$d_{\text{esu}} = 3.7 \times 10^{-15} d_{\text{mks}}$$

where ε_0 is included in the coefficient d.

b. Calculate in mks units

$$P_{ne} = 2d_{14}E^2 \quad \text{for} \quad E = 10^8 \text{ N/C}$$

Solution

(a)

$$P_a = 1.7\times10^{-41} \ \frac{\text{mC}^2}{\text{N}} \cdot 5\times10^5 \ \frac{\text{N}}{\text{C}}$$

$$= 8.5\times10^{-36} \ \text{m}\cdot\text{C}$$

(b)

$$P_{ne} = 2\cdot d_{14}\cdot E^2$$

$$= 2\cdot(1.6\times10^{-9})_{\text{esu}} \ (10^{16}) \ \left(\frac{\text{N}}{\text{C}}\right)^2$$

$$= 2\cdot(1.6\times10^{-9}) \ (3.7\times10^{-15})_{\text{mKs}} \ 10^{16} \left(\frac{\text{N}}{\text{C}}\right)^2$$

$$= 2\cdot(5.9\times10^{-9-15})_{\text{mKs}} \ 10^{16} \ \left(\frac{\text{N}}{\text{C}}\right)^2$$

$$= 2\cdot(5.9\times10^{-24})_{\text{mKs}} \ 10^{16} \ \left(\frac{\text{N}}{\text{C}}\right)^2$$

$$= 11.8\times10^{-8} \ \text{C}\cdot\text{m}$$

8. *Review of the "pancake model".*

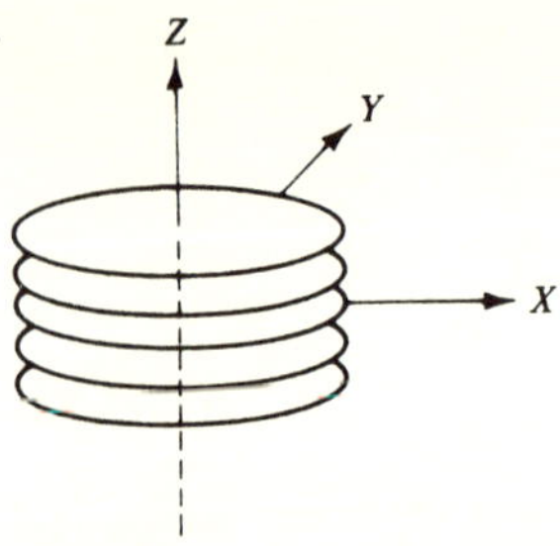

(a)

For waves vibrating in the X, Y plane, the ordinary refractive index applies; for waves vibrating parallel to the Z axis, the extraordinary index applies. For the propagation vectors, we have

$$k_X \text{ (polarization of wave } \parallel \text{ to } Y) \qquad k_X \to \text{ ordinary} \to k_{Xo}$$
$$k_X \text{ (polarization of wave } \parallel \text{ to } Z) \qquad k_X \to \text{ extraordinary} \to k_{Xeo}$$
$$k_Z \text{ (polarization of wave } \parallel \text{ to } Y) \qquad k_Z \to \text{ ordinary} \to k_{Zo}$$
$$k_Z \text{ (polarization of wave } \parallel \text{ to } X) \qquad k_Z \to \text{ ordinary} \to k_{Zo}$$

For a negative uniaxial crystal for the k-ellipsoid ($n_e < n_o, k_e < k_0$) we have the following k surfaces:

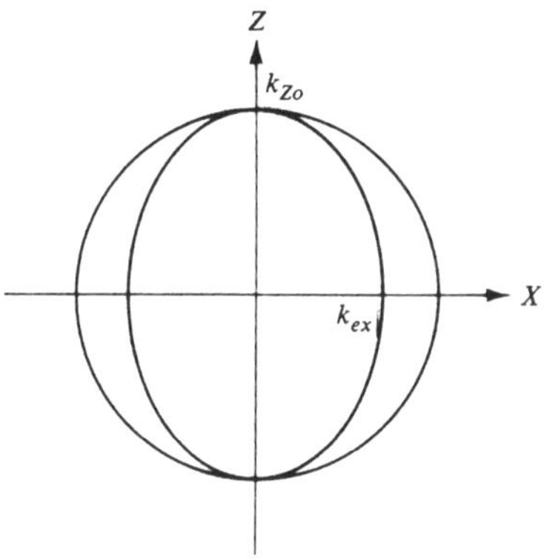

(b)

This may also be drawn as an index ellipsoid:

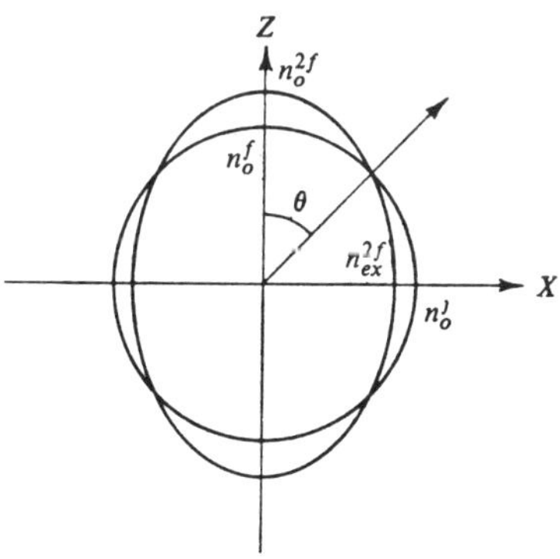

(c)

We have superimposed the index ellipsoid for the second harmonic, which has values $n_0(2f_o)$ in the Z direction and $n_{ex}(2f_o)$ in the X direction. Index matching occurs at angles θ_m, which may be calculated from

$$\frac{\cos^2 \theta_m}{[n_o(2f_o)]^2} + \frac{\sin^2 \theta_m}{[n_{ex}(2f_o)]^2} = \frac{1}{[n_o(f_o)]^2}$$

a. Show that we have

$$\sin^2\theta_m = \frac{1/[n_o(f_o)]^2 - 1/[n_o(2f_o)]^2}{1/[n_{ex}(2f_o)]^2 - 1/[n_o(2f_o)]^2}$$

b. For KDP, we have for $\lambda = 6940$Å

$$n_{ex}(f_o) = 1.466$$
$$n_o(f_o) = 1.506$$

and for the second harmonic

$$n_{ex}(2f_o) = 1.487$$
$$n_o(2f_o) = 1.534$$

Show that $\theta_m = 50.4°$.

$$n_o^{2f} \equiv X$$
$$n_e^{2f} \equiv Y$$
$$n_o^{f} \equiv Z$$

Solution

(a)

$$\frac{\cos^2\theta}{X^2} + \frac{\sin^2\theta}{Y^2} = \frac{1}{Z^2}$$

$$\frac{1-\sin^2\theta}{X^2} + \frac{\sin^2\theta}{Y^2} = \frac{1}{Z^2}$$

$$\frac{1}{X^2} - \frac{\sin^2\theta}{X^2} + \frac{\sin^2\theta}{Y^2} = \frac{1}{Z^2}$$

$$\sin^2\theta\left(\frac{1}{Y^2} - \frac{1}{X^2}\right) = \frac{1}{Z^2} - \frac{1}{X^2}$$

$$\sin^2\theta = \frac{\frac{1}{Z^2} - \frac{1}{X^2}}{\frac{1}{Y^2} - \frac{1}{X^2}} = \frac{\frac{X^2}{Z^2} - 1}{\frac{X^2}{Y^2} - 1}$$

$$\sin^2\theta = \frac{\frac{1}{(n_o^f)^2} - \frac{1}{(n_o^{2f})^2}}{\frac{1}{(n_e^{2f})^2} - \frac{1}{(n_o^{2f})^2}} = \frac{X^2 - X^2}{\frac{X^2Z^2}{Y^2} - Z^2}$$

$$= \frac{X^2Y^2 - Z^2Y^2}{X^2Z^2 - Z^2Y^2}$$

$$= \frac{\left(\frac{n_o^{2f}}{n_o^f}\right)^2 - 1}{\left(\frac{n_o^{2f}}{n_e^{2f}}\right)^2 - 1}$$

(b)

$$\sin^2\theta = \frac{\left(\frac{1.534}{1.506}\right)^2 - 1}{\left(\frac{1.534}{1.487}\right)^2 - 1} = \frac{0.0375}{0.0642} = 0.58$$

$$\underline{\underline{50.4^\circ}}$$